José Laerte Cecílio Tetila

The Environment

José Laerte Cecílio Tetila

The Environment

Mato Grosso do Sul in Focus

ScienciaScripts

Imprint
Any brand names and product names mentioned in this book are subject to trademark, brand or patent protection and are trademarks or registered trademarks of their respective holders. The use of brand names, product names, common names, trade names, product descriptions etc. even without a particular marking in this work is in no way to be construed to mean that such names may be regarded as unrestricted in respect of trademark and brand protection legislation and could thus be used by anyone.

Cover image: www.ingimage.com

This book is a translation from the original published under ISBN 978-3-330-76342-5.

Publisher:
Sciencia Scripts
is a trademark of
Dodo Books Indian Ocean Ltd. and OmniScriptum S.R.L publishing group

120 High Road, East Finchley, London, N2 9ED, United Kingdom
Str. Armeneasca 28/1, office 1, Chisinau MD-2012, Republic of Moldova, Europe
Managing Directors: Ieva Konstantinova, Victoria Ursu
info@omniscriptum.com

Printed at: see last page
ISBN: 978-620-8-40768-1

To my children André Luís, Andyara and Andyane.

To my grandchildren Ana Clara, Arthur, Theodoro, Lívia and Francisco.

ACKNOWLEDGEMENTS

To my partner Zonir, who lovingly read and revised the text, making important suggestions.

To the journalists Luiz Taques, Marcelo Pereira, Dalva Gonçalves, Helton Costa, Wender Carbonari and also Eva Garcia Dantas, Ângela Maria Freitas de Souza, Jovina Nevoleti Correira who read the articles and/or helped organise the material.

To Prof Dr Maria Eugênia Carvalho do Amaral for her suggestions.

ABOUT THE AUTHOR

Laerte Tetila was born in Santo Anastácio (SP) on 7 July 1947. He is married to Zonir Freitas Tetila and has a son, two daughters, three grandsons and two granddaughters. In 1975, he graduated in Geography from the then Faculdade Dom Aquino, in Campo Grande - now the Universidade Católica Dom Bosco (UCDB). He specialised in Physical Geography at the Dom Bosco College in Santa Rosa (RS) and holds a master's degree in Physical Geography from the University of São Paulo (USP).

He was a teacher at the "Castro Alves" State School, and at the "Presidente Vargas" and "Menodora Fialho de Figueiredo" schools in Dourados (1968 to 1978), and also taught at the Dourados Pedagogical Centre (UEMT) from 1975 to 1978.

He was a professor of Geography at CEUD/UFMS from 1979 to 1996, where he held the positions of Course Coordinator, Head of Department and was a member of the University Council. He published the book *Marçal Tupã I* - um *guarani que não se cala* (*Marçal Tupã I* - a *Guarani who won't shut up*) (Ed. UFMS, 1994). He also co-authored, with Wilson Biasotto, "O *Movimento reivindicatório do Magistério Público Estadual de Mato Grosso do Sul: 1978/1988" (Ed. UFMS).*

Tetila was a councillor in Dourados (1993-1996 and 1997-1998), state deputy (1999-2000) and mayor of Dourados for two terms (2001-2004 and 2005-2008). In 2011, he was once again elected to the state legislature.

In the Legislative Assembly, Tetila worked in defence of education, the environment, human rights, family farming and indigenous communities. President of the Labour, Citizenship and Human Rights Commission, he was the author of the law that created the State Fund for the Acquisition of Indigenous Lands, Fepati.

PREFACE

FIGHTING FOR ENVIRONMENTAL ISSUES AND MUCH MORE

Laerte Tetila is, above all because he is a teacher and a politician, a fighter. And I had the privilege of having him as a work colleague in the 1970s, at the embryonic UFGD Pedagogical Centre in Dourados at the State University of Mato Grosso, where we shared expectations and dreams inside and outside the classroom, including starting a master's degree, a daring undertaking at the time.

Almost 40 years later, here I am witnessing yet another of Laerte's daring acts, disguised to this day by a contumacious serenity: showing the reality of Mato Grosso do Sul's development, centred on profit maximisation, and its unquestionable "price" - disregard for the environment and poor quality of life - and proposing alternatives based on the Culture of Peace.

In this book, the teacher speaks louder than the politician and the politician shows what he has learnt from the teacher. Each article is a didactic lesson, full of data and quotes on environmental issues. A lesson in the essential role of riparian forests, the Guarani aquifer, biodiversity, agroecology, ecological corridors, windbreaks, preservation units (UPAs), clean energy sources and sustainable development, as well as the advantages of a technological agrarian reform. Nor are the articles oblivious to the irreversible damage caused by hydroelectric power stations on the Pantanal's edge plateaus, as well as the consequences of rural and urban fires, water pollution, erosion, deforestation and the contamination of water and soil by agrochemicals; in addition to the providential warning, so little clarified by public health bodies, about the dangers of urban smoke.

It's almost impossible not to get involved with the passion with which Laerte writes about each environmental issue. At times, the professor uses images on the threshold of science and poetry:

"Protoplasm, for example, which is found in all animals and plants, but

> also in soil, air, water and, incredibly, even in the hardest rocks, is the basis of human life. And since nature is so present in us, with all its elements, there's no way we can't be nature too."

At other times, the text is blunt:

> "The problem is that deforestation is linked to a linear, growing economic model based on competition. A model that commodifies nature, decimates vital processes and increases environmental pressure at an astonishing and unprecedented rate."

And let's emphasise Laerte Tetila's political vision, which is so evident in the article "Mata de Dourados, uma devastação implacável":

> "The Dourados Forest, which had been untouched until 1943, had around 300,000 hectares cleared for the establishment of the successful National Agricultural Colony. So far, so good. From 1970 onwards, however, with the expansion of the agro-export model, what we saw was a real environmental disaster. The scythe and the axe were replaced by chainsaws and tractors with blades, and fires increased alarmingly. Hence the escalation of gigantic deforestation, in flagrant disregard for biodiversity, ecological sustainability and the Brazilian Forest Code."

However, let's be clear: each article has its own context and time (providentially indicated by the date it was written), even though there are timeless issues in the author's approach. With a didactic explanation, Laerte translates the progress made in the environmental struggle, citing, among others, the Stockholm Conference (1972), Rio92 (1992), the Dublin Conference (1993), the Kyoto Protocol (1997) and Rio+20 (2012).

Even with the frustrating outcome of Rio+20 - the United Nations Conference on Sustainable Development - which produced "an insouciant document, fearful in its intentions, without targets and without money" (as the Senior Vice-President of the Americas Division of the NGO Conservation International, Prof Dr Fabio Scarano, declared in June 2012), Laerte Tetila didn't give up and, at the end of the same year, wrote "Sustainability is the word of the century". In this article, he cites the United Nations' Inclusive Wealth Index (IRI) and criticises the use of Gross Domestic Product (GDP) as the main indicator for assessing a country's development, increasing the economic obsession with "growth at any cost". After all, as he states in another article from March 2013:

> "Growth, although necessary, is not enough, as it does not guarantee development, which in turn is a process of economic, social, cultural and environmental transformation that implies an improvement in the well-being of society as a whole. A rise in everyone's standard of living."

In this and other texts, the author makes it clear that sustainable development is the policy of common sense, with obvious and natural respect for human rights. And before I go on commenting on each article, join us now in the company of professor and (fortunately) politician, Laerte Tetila.

A good lesson and good questions for you, reader.

On 8 March 2014.

Maria Eugênia Carvalho do Amaral Master's and PhD in Ecology from UNICAMP Retired full professor at UFMS

PRESENTATION

We're killing planet Earth like we kill ants. Unfortunately, attacks on nature continue to be more intense than ever. Minerals are being depleted, fires are increasing, soils are degrading, deserts are advancing, the atmosphere is becoming disfigured, water is rotting, fauna and flora are being suppressed, agricultural areas are being poisoned and food is being contaminated.

This collection of articles, focussing on Mato Grosso do Sul, brings together a discussion of challenges for the future of the environment and life. Most of them have been published in newspapers or on websites and, now gathered together, are intended to help people rethink and increasingly contribute to preservation initiatives, towards building the longed-for environmental sustainability and social well-being.

CONTENTS

CHAPTER 1

THE GEME PLANET

Damage to the environment continues to be more intense than ever. Environmental degradation is rampant, out of control and out of bounds. Just look at the ozone layer being eroded, the greenhouse effect increasing, temperatures rising, the atmosphere being disfigured, glaciers melting, waters rotting and deserts advancing. And what about fossil fuels, pesticides, radioactive waste, population growth and consumerism?

The pressure on the environment has become frightening and the planet is groaning like a living organism, just like in the Gaia theory. We have reached the imponderable. The rate of exploitation of natural resources already exceeds the rate of regeneration. It's incredible, but planet Earth has already been dilapidated 30 per cent more than it can bear, according to a warning from the world's own scientific community.

This is the environmental liability. A liability that has fallen on the Biosphere (the support of life), the price of which humanity has been paying with a brutal and unprecedented loss of biodiversity.

It's distressing to realise that well over a hundred species a day, including plants and animals, are being wiped off the face of the earth forever, despite being the fruit of millennia of organic evolution. This is extinction, which will consume no less than half of the Biosphere this century.

Cutting greenhouse gas emissions has been a failure. Climatology centres around the world have revealed the seriousness of the problem. Global warming continues unabated and one of its many consequences is desertification, which has already affected more than a hundred countries, according to the UN, and which has produced around 500 million environmental refugees, to say the least.

Although the environmental crisis has become a truism, few take it seriously. Even with the planet showing visible, sensitive and measurable signs of exhaustion, even with all the indicators demonstrating the extent of the damage, the war for the appropriation of the remaining natural resources continues to be more intense and unbridled than ever, and worse:

even fuelled by modern, state-of-the-art technologies.

It's the rhythm dictated by the well-established pattern of production and consumption, focused on maximising profit, which continues indifferently, caring little for the environmental damage and the future of humanity.

Not even the world's mega-events, exposing the seriousness of the crisis with scientific clarity, such as the Stockholm Conference (1972), the Earth Charter (1987), the Kyoto Protocol (1997), Eco-Rio (1992), the Johannesburg World Summit (2002) and the famous Agenda 21, have been enough to contain the momentum of the binomial productivism-consumerism and its deleterious and devastating effects.

Sustainability, a magnificent concept that aims to satisfy present needs without compromising future generations, consolidated at Eco-92, now unanimous and universal, has unfortunately only increased the frustration of the UN and the world scientific community, because the greater the accumulation of information and warnings, the greater the effort to raise awareness, the more we have squandered and destroyed. On the other hand, it would be an injustice to want to blame all of humanity for the worsening crisis, without bringing the protagonists of productivism and consumerism to light. And it's not to be doubted that these medallions, because they occupy the top of the economic pyramid, don't even feel like members of humanity.

Unfortunately, initiatives in favour of the environment have been timid and only occasional. Reconciliation and a new pact with nature are still words in the wind. The concept of sustainable development, although practised seriously by some, has already been appropriated and transformed into a good business opportunity through misleading propaganda. Hence the almost unanimous scepticism towards Rio+20. That's why Lampeduza's old maxim has never been more relevant: let everything change so that everything remains as it is.

THE ENVIRONMENTAL STRUGGLE IS NOT MERE LYRICISM

For some time now, human beings have shown enormous difficulty in coexisting with the environment. Throughout history, society has distanced itself from nature as if it wasn't part of it, as if it didn't have to preserve it. Today, when we analyse the relationship between the two (society and nature), we soon realise how unfavourable this relationship has been for nature.

There are still those who degrade as if it were something natural. Meanwhile, the pressure on the environment, resulting in huge environmental liabilities, continues to cancel out vital processes of the utmost importance.

As a result, components of the biosphere have been eliminated in geometric progression. And when more than 100 species (animals and plants) are wiped out forever, every day, by extinction, we can only be looking at exploitation that goes far beyond sustainability. Nature is being squandered year after year.

There is no denying the evidence of an unprecedented environmental crisis. A crisis that certainly comes from no other source than the infamous Cartesian rationalism, which still manages to reach us with its old anthropocentric postulate, where the relationship between society and nature is nothing other than one of superiority and domination.

Because of this, in the imagination of many people, human beings must still see nature as a servant, in other words, as something that belongs to them to be used, dominated and transformed as they please. And if it is degraded, it must be abandoned, left behind. It's unfortunate, but this is the view that still serves as the basis for justifying the unbridled exploitation of natural resources, as if they were infinite.

However, although there are still those who are charmed by such a stance, researchers and scholars around the world are not in that boat. In fact, many never have been, because it's crystal clear that this mechanical, dichotomous relationship between society and nature has never been a good one.

Today, in the new biocentric vision, nature, in turn, is intrinsic to the human being. Protoplasm, for example, which is found in all animals and plants, but also in soil, air, water and, incredibly, even in the most solid rocks, is the basis of human life. And since nature is so present in us, with all its elements, there's no way we can't be nature too. Hence the focus of this new vision is centred on defending a human-nature relationship of balance, cooperation, harmony, synergy, reverence, interactivity and sustainability. In short, it is an affirmation of the value of nature and life as a fundamental requirement for our own survival.

In this holistic, environmentalist way, there's no way we can lose touch with our biological and natural basis, so divinely planned. In this new way, the environmental struggle is no longer mere lyricism.

Let's face it, we're losing a large part of our biodiversity. The trend must be stopped. It's crystal clear that the break with nature is the path to self-destruction. A path of no return. Initiatives focussed on modern environmentalism are trying to build a new horizon - a transition to a new, interactive and sustainable development model.

CHAPTER 2

BRAZIL, WORLD LEADER IN DEFORESTATION

It's threefold that the gentry speaks and the human race doesn't listen (Vitor Hugo)

Matches never stopped smouldering and chainsaws never stopped snarling. And there goes more than half of the planet's vegetation cover. The devastation, despite the worldwide fight in defence of the environment, has only grown. Destruction, especially of forests, has increased by leaps and bounds, and Brazil is, unfortunately, number *one in the ranking of forest destruction*[1] , and is therefore the world leader in terms of forest devastation.

It's not that Brazil lacks environmental legislation and enforcement. Despite the heavy fines levied on thousands of landowners, the impression is that illegal deforestation will only stop when the last tree in this country is felled. And this is not alarmism. The figures speak for themselves. In Mato Grosso do Sul, for example, IBAMA - the Brazilian Institute for the Environment and Renewable Natural Resources - *issued 283 infraction notices in just one year (2010), totalling R$ 51 million in fines, which is not why illegal deforestation has* stopped .[2]

And it's been like this all over Brazil. According to Wagner Cerqueira, *approximately 20,000 square kilometres of native vegetation are deforested as a result of fires*[3] .

It's unfortunate, but you can already imagine that our country is bidding farewell to its forests and its rich biodiversity.

The problem is that deforestation is linked to a linear, growing economic model based on competition. A model that commodifies nature, decimates vital processes and increases environmental pressure at an astonishing and unprecedented rate.

1 ALENCASTRO, Catarina. Brazil is first in the ranking of forest clearing. O Globo newspaper, 24.09.2014 - updated on 24.09.2016.
2 ALMOAS, Jorge. IBAMA imposed R$ 51 million in fines in Mato Grosso do Sul in 2010. Campo Grande News, 21 December 2010.
3 CERQUEIRA, Wagner. Deforestation in Brazil. Available at mundoeducacao.uol.com.br/geografia/desmatamento-no-brasil.htm.

Whether legal or not, deforestation doesn't stop and the profound and irreversible impacts are there, such as the loss of biodiversity, damage to soil and water resources, carbon dioxide emissions, climate change, the brutal extinction of fauna and flora and, finally, desertification, the process of which is already underway in the Northeastern Sertão, the Cerrados of Tocantins, northern Minas Gerais and Mato Grosso do Sul, the Pampas, to say the least. And there's no forestry code that can fix it. The anti-ecological lobby has been heavy.

For Greenpeace, the text approved at Rio+20 was an epic failure. Once again, they managed to separate ecology from economics. Once again, the economic model won. The destructive, out-of-control escalation won. The Earth Charter (UN), in stating that our common destiny calls us to seek a new beginning, has struck at the heart of the problem. It remains to be seen whether society has the will to react, albeit belatedly, and set out in search of a new economic model, with a new consciousness and based on sustainability.

NO NEED TO CUT DOWN ANY TREES

The idea that we should cut down our forests because the first world countries have done so, or that we need to deforest in order to develop, needs urgent repair. The mistakes of the past do not justify those of the present.

Those countries did indeed make many mistakes, but they had to learn from them. So much so that, according to Thiago Fonseca Morello, forestry has been growing since the 19th century in Scotland, Denmark, France, Switzerland, Germany, South Korea, England and the United States, countries that have adopted mechanisms to encourage reforestation and the regeneration of native forests.[4]

Recognising the immense value of forests, these countries have adopted what is known as low-impact (sustainable) forest management, which turns forests into a precious source of employment and income without destroying them, all the more so given their importance in protecting the climate, the watersheds, biodiversity and human life itself.

In Japan, for example, despite the high population density, the increase in forest cover has

4 MORELLO, Thiago Fonseca. Deforestation is not development. Planeta Magazine, São Paulo, 14 December 2012.

been impressive. According to Morello, 69 per cent of Japanese territory has already been reforested, thanks to the establishment of biodiverse forest plantations. In France too, according to Morello, the forest area has not been reduced since the mid-19th century; on the contrary, it has increased, and France is now self-sufficient in wood.

Against the backdrop of history, forests continue to shrink in emerging countries, especially tropical ones, and Brazil is among those with the highest deforestation rates. But despite the loss of biodiversity and the tremendous environmental liabilities that are accumulating in our country, everything indicates that the concept of sustainable development is already beginning to bear fruit. Let's see: for Eduardo Riedel, president of the Mato Grosso do Sul Agriculture Federation (Famasul), "those who deforest are not represented by us, and they have to be punished. There is room for production to grow without the risk of environmental degradation, and it is estimated that in Mato Grosso do Sul there is room for environmental degradation.

Grosso do Sul, has an area of nine million degraded hectares, and it is with the use of *these areas that production should expand*[5]

Brazil already has 200 million degraded pastures. According to Paulo Barreto, from the Amazon Institute for Man and the Environment (Imazon), *it is possible to double or triple Brazil's agricultural production without cutting down trees. To do this, according to Embrapa, it would be necessary to increase the productivity of already deforested areas*[6] . In fact, only with crop-livestock integration will the desired grain and meat production be achieved, without further sacrifices to plant formations. The application of technologies for the recovery of degraded areas, including subsidised credit lines, is already available, with good possibilities for increasing production without the need for new *clearing*. If you think about it, the nine million hectares of degraded, under-utilised land in the south of Mato Grosso, even close to the asphalt network, signal the size of the contradiction of further deforestation.

5 Production should grow using degraded areas. Interview with Eduardo Riedel for Correio do Estado. Campo Grande, MS, 30.Sep.2011.

6 BARRETO, Paulo. Deforestation is unnecessary. Folha de São Paulo - Opinion, 27 June 2009.

NOTHING REPLACES RIPARIAN FORESTS

Anyone who flies over the territory of Mato Grosso can see, along its rivers and streams, the riparian forests being transformed, with rare exceptions, into simple strands of vegetation. Anyone who sails on the state's rivers can also see the narrowing and thinning of this strip of vegetation, if not stretches where it has been completely devastated.

In Mato Grosso do Sul, despite the enormous environmental liabilities that can be seen in all of its biomes, the focus of pressure is unfortunately now falling on these forests, despite the extent of the degradation they have already suffered. But given the remarkable water potential of this state - just look at its extraordinary wealth of rivers, lakes, streams and springs - this shouldn't be happening.

On the contrary, instead of the pressure to reduce or even eliminate them, we should follow the example of neighbouring Paraná, where, since 2003, more than 100 million seedlings of native species have been planted along its rivers and other water sources through the Riparian Forests programme[7] , following a trend already present in some countries.

In Europe and North America, millions of dollars are being spent on restoring their riparian forests in an attempt to save their rivers. The UN even has a specific programme for restoring this type of forest. Nature is demanding back what it has lost.

Nothing replaces riparian forests as a factor in protecting water bodies. These forests were originally dense strips of vegetation, rich in leafy trees that reached up to 30 metres.

They were home to a rich variety of plants and animals, the fruit of millions of years of organic evolution. And, in addition to their extremely high biodiversity, they were also legitimate providers of environmental services, fulfilling an important role as physical barriers that reduced the force of the floods that reached the rivers. They also functioned as hydro-regulating sponges, retaining rainwater and gradually releasing it in order to recharge rivers and springs, thus reducing the damaging effects of floods and ebb tides. The tangle of roots that reached the banks stabilised the ravines to prevent landslides and siltation, phenomena

7 Programme Riparian Programme Programme State of Paraná. Available at http/www.meioambientre.pr.gov.br/modules/conteúdo/conteúdo phd?con

that now affect practically all rivers in the state of Mato Grosso do Sul.

It's obvious that, in the areas where they still exist, they continue to fulfil their functions. Even in the face of the hitherto unimaginable deforestation of the state's biomes, especially the woodlands and savannahs, riparian forests are still functioning as true refuges for terrestrial fauna and birdlife.

Although degraded and disfigured as they are, they still insist on their noble mission of serving as ecological corridors, where wildlife moves, shelters, feeds, reproduces and spreads seeds.

Riparian forests also serve as natural filters that retain chemical fertilisers and pesticides, reducing river pollution and consequently improving the quality of the water that is often used for urban supply.

DEFORESTATION IN THE AMAZON AFFECTS MATO GROSSO DO SUL, OTHER STATES AND NEIGHBOURING COUNTRIES.

For the people of Mato Grosso do Sul, the Amazon rainforest should mean much more than a huge green spot on the map, given its extraordinary importance for the state. Heated by the equatorial sun and due to the intense transpiration of the plants, this forest, with the help of evaporation from the rivers (the largest hydrographic basin in the world), ends up producing a gigantic mass of humid air, with intense rainfall, which is essential for biodiversity and farming in Mato Grosso. However, this huge mass of humidity and rain might not reach Mato Grosso do Sul. Its certain destination would be the Pacific Ocean, were it not for the providential wall of the Andes with its four thousand metres of altitude. A wall that, by barring it, redirects it towards the territory of Mato Grosso do Sul to irrigate it from October to March with copious rainfall, providing it with huge crops and pastures, just as it does with the states of Mato Grosso, Paraná, Santa Catarina, Rio Grande do Sul and neighbouring countries such as Paraguay and Argentina.

At the beginning of the 1980s, when I was a postgraduate student at USP, information that reached us from the National Institute for Aerospace Research (INPE) already warned that the deforestation of the Amazon rainforest would cause a serious reduction in rainfall in the Centre-South of the country and that the territory of Mato Grosso do Sul would be one of the worst affected. Nowadays, however, much more advanced research, based on satellite images, has only reaffirmed this thesis in a much more scientific way.

The destruction of forests is a serious matter. Despite all the environmental awareness, plant formations in general continue to be the biggest victims of progress. In the Amazon, with the use of fire and modern technology, vast tracts of forest are soon converted into space for soya and cattle. Deforestation sometimes slows down, but never stops. To give you an idea, according to INPE, between August 2010 and July 2011 alone, 6,418 km were devastated.[8]

Faced with this reality, the disastrous consequences for the economy and for life in the state of Mato Grosso can no longer be left to speculation, because the more deforestation there is in the Amazon, the longer the droughts will last in the state, which, in time, could face the agonising problem of desertification.

The surprising drought of 2005 in the Amazon region, which caused irreparable damage to rural production in the interior of the country, was a worrying sign of what not to do with the Amazon rainforest. A sign of what could happen when the Amazonian air mass, this suspended water giant, is prevented by deforestation from forming and moving through these plagues.

Our Mato Grosso do Sul needs to put the problem of deforestation in the Amazon on the national environmental agenda and take a preventive stance, because just as Egypt is a gift from the Nile, Mato Grosso do Sul is certainly a gift from the Amazon rainforest. A gigantic gift, with its billions of litres of water suspended in the air. A gift full of humidity, which condenses and precipitates copiously to fill our rivers, lakes and reservoirs and make crops and pastures sprout and life flourish with all its intensity. A gift that, in geography, is known as the Continental Equatorial Air Mass.

DEFORESTATION, WINDS AND SOIL DRYING IN MATO GROSSO DO SUL

It's true that in the agricultural areas of Mato Grosso do Sul, the winds have been removing moisture from the soil - and increasingly causing it to dry out. This is an environmental problem that has caught the attention of agro-climatologists around the world.

8 INPE News, 05.Jun.2012, Available at: HTTP:/www.cpetec.inpe.br/noticias/noticias/128679

Through the winds, soils dry out as they speed up. And this is worrying, because the surface winds (lower troposphere), released by deforestation, are gaining speed and, in the process, larger and larger spaces are being exposed to wind action.

In this state, huge geographical areas, originally covered by the Cerrado and the Atlantic Rainforest (Mata de Dourados), have been devastated far beyond the 80 per cent stipulated in the Brazilian Forest Code, in order to make room for plantations and pastures. Had the legal reserves of 20 per cent been maintained on all properties, we would certainly have a veritable mosaic of vegetation on the landscape today which, among many benefits, would serve as a buffer to dampen the speed of the winds and thus improve the water availability of the soil.

Deforestation has been and continues to be a devastating phenomenon. In the municipality of Dourados, for example, 97 per cent of the forest cover has already come down[9] , and this is the tone of the whole of Greater Dourados and the Ivinhema Valley, where the Atlantic Forest (Mata de Dourados) used to reign supreme.

And you don't even have to fly over these areas or the Cerrado to realise the scale of the problem. Accelerated deforestation has not yet stopped in these regions, despite the fact that it is nearing its terminal phase.

And it's not just an optical illusion. The brutal escalation of deforestation is explicit in the landscape, so that the last fragments of forest and savannah, with very few exceptions, are being burnt and destroyed. In this way, generalised deforestation has resulted in what is now known as the "*descampados*". These areas, although occupied by a veritable agricultural granary, expose the monotony of a landscape totally unsheltered by surface winds.

Winds that, in addition to the open fields, have another great ally in their favour: topography. Thus, open moorland areas, with some of the smoothest topography in the world, make up the binomial that only exacerbates the problem, because, acting freely, these winds (from the lower troposphere) have started to remove as much moisture as they can from the soil, as if collecting the bill for deforestation that has gone far beyond what nature can bear.

In fact, the increase in wind speed in these ancient areas of woodland and savannah in the south of Mato Grosso is nothing more than the manifestation of a serious imbalance which,

9 Atlas of Atlantic Forest Remnants, results by municipality. SOS Mata Atlântica Foundation, May 2009.

without restorative measures, will only tend to increase the negative impacts, such as droughts and summers.

In view of the extent of deforestation in the Cerrado and Mata de Dourados regions and the environmental problems that jeopardise the future, what is hoped for is the emergence of initiatives and mitigating measures that will gradually ease the problem. No-till farming is already an important initiative. Recovering the 20 per cent of legal reserves with native species would create mosaics of vegetation, which are fundamental as windbreaks.

The implementation of windbreaks, formed by rows of trees and shrubs arranged perpendicular to the prevailing winds, would also be an important measure. A measure that goes back a long way, but which has received little attention, despite Embrapa's good research on the subject.

Recognising its importance, countries such as the United States, Denmark and Russia have even implemented windbreaks on a national scale.

FIRES AND THE MARTYRDOM OF SOUTHERN MATO GROSSO'S FAUNA GROSSENSE

As soon as the dry season arrives, the ravages of wildfires can already be seen everywhere. And it's heartbreaking to realise that, in Mato Grosso do Sul, thousands of wild animals are burned alive during the dry season.

The territory of Mato Grosso do Sul, which in the past was noted for its beautiful and enchanting fauna, today there is no denying society's indifference to it, just by looking at what it has left.

And this is not alarmism. The crisis experienced by the animals that make up the fauna of this state is evident everywhere and in every light. It's a fauna under siege. The fires have mercilessly advanced over tree, shrub and herbaceous formations and even over environmental protection areas.

The indiscriminate use of fire in rural areas has become a serious issue and, according to Bárbara Cavalcante, Mato Grosso do Sul is already the state with the highest number of fires in the country.[10]

The scenes of burning are those of a holocaust. Fire, a powerful communicant, martyrs the earth. Its effect is deleterious, devastating. It destroys litters, kills baby chicks in their infancy and mortally wounds all forms of life.

In a fire, the temperature can exceed 800° centigrade and the flames leave painful marks. Animals are burnt alive, if not injured or poisoned. Some of them are volatilised without leaving the slightest trace. The injured flee, aimlessly, to die of infection. The survivors, expelled from their habitat, face a lack of food and attacks from predators in strange places. The impact is cruel.

The use of fire, which agonises wildlife and destroys biodiversity, is a harmful, outdated technique from the Neolithic Age. And even though burning is an environmental crime (Law 9605/98) and research shows that there are far more advanced techniques for improving productivity than fire.

Even so, there are still the archaic ones, the pyromaniacs, those who set fires without the slightest compassion for animals.

Burning, which is universally considered to be one of the main causes of species extinction, seems to have reached its apogee in the south of Mato Grosso. Here, the burning of natural environments is unbelievable. Fire continues to mow down flocks of defenceless animals, despite all the education available and all the awareness campaigns.

Many people don't want to understand that nature was divinely planned and that the animals that burn are the result of millennia of organic evolution. That although human beings have reached the moon, they will never make a tuiuiú, a maned wolf, a hyacinth macaw or a songbird. That you can't rebuild genetic material lost to extinction. They don't want to understand that, in nature, the loss of a species means a subtraction in the machinery of life. An eternal loss for biodiversity.

So what is the future for this fauna? We need to think about changes, and not so much through laws, but through minds and hearts. It's not possible that, in the middle of the third

10 Cavalcante, Bárbara. Mato Grosso do Sul has the highest number of fires in the country. Correio do Estado - Weather. Campo Grande - MS, 12 January 2017.

millennium, fire is still being used as a farming technique. It is impossible that there is no compassion for the little wildlife that remains in these fields, forests and wetlands.

IVINHEMA RIVER FLOODPLAIN PARK AND FOREST FIRES

The idea of creating park areas first came from ancient China. With its ancient culture, that country preserved some areas as early as antiquity, but the idea gained momentum in 1872 with the creation of Yellowstone Park in the USA. The Várzeas do Ivinhema Park, inaugurated in 1998, became the first to be set up in Mato Grosso do Sul. With 73,300 hectares, it was acquired by Companhia Energética de São Paulo (CESP) and transferred to the state of Mato Grosso do Sul as compensation for the construction of the Sergio Motta Dam in Porto Primavera, on the Paraná River.

On a visit to the site, accompanied by Captain Carlos Magno da Silva, of the MS Environmental Police, and other police officers under his command, it was possible to see that this is a preservation area that impresses with its size and the structure installed by CESP. This structure is made up of housing for the Environmental Police, accommodation for visitors and/or researchers, a research centre and an observation tower - from where you have a wide view of the local biodiversity, such as the Atlantic Forest, Cerrado, varjões and riparian forests.

By boat from the Environmental Police, we travelled down to very close to where the waters of the Ivinhema River meet the Paraná River. A little further south, there is another park, Ilha Grande, where an ecological corridor is expected to be set up, of fundamental importance for a broad integration between the fauna and flora of both parks. Once the two parks are joined, an immense ecological space will be guaranteed in the far south of the state, which will certainly become a benchmark as a public environmental preservation area.

It should be emphasised that where the Paraná River meets Ilha Grande, it has become a unique place in the immensity of this river that does not suffer from flooding from dams, a fact that gives it a special condition, namely that of becoming the only large fish spawning area (natural breeding ground) along the whole of the great Paraná River. From there, after the spawning period (piracema), the fry come downstream to populate what is one of the most dammed rivers in the world.

The problem is that, time and again, the park has been disfigured by the action of fire, because before it became a park area, it was occupied by livestock. With the removal of the cattle, however, more than 50,000 hectares of brachiaria and/or colonião have been left at the mercy of their vital cycles: they are born, grow, dry out and release their seeds. When you least expect it, kilometre-long fire fronts appear, raising gigantic, uncontrollable flames. Hundreds of nests, eggs, chicks, reptiles, mammals and everything else are devoured.

However, this problem can be overcome through self-management, because if the compensation from CESP and the resources from the MoH are not enough, the park itself, in our opinion, is in a position to generate its own income, since, apart from the rivers, lakes and varjões, there are still more than 50,000 hectares of area left to be used through economically profitable and ecologically correct management, which could even serve as a model for the other park areas in the MoH.

Government action, which brings together environmentalists from various fields, is essential for the formatting of an up-to-date, modern management plan that identifies solutions for both the fires and the restoration of these biomes, given the extraordinary potential for tourism and scientific research present in the area.

CHAPTER 3

URBAN FIRES AND PUBLIC HEALTH

Winter, low humidity and, of course, dry vegetation in both the countryside and the cities. And you don't even need satellite images during this season of the year to spot urban and peri-urban fire outbreaks, because in the wastelands, taken over by the dry mace, the fires resurface relentlessly with their flames, generating the most serious environmental impacts.

The sum of all the fires ends up releasing tonnes of toxic gases, such as carbon dioxide and methane. And in winter, the cold, dense air, pressing down, only makes it more difficult to dissipate these gases, which, from being visible (smoke, when burning), become invisible, in the form of millions of toxic particles, forming an almost homogeneous layer in the lower troposphere, which is inhaled by everyone, without distinction, becoming a serious public health problem.

To give you an idea, at this time of year the amount of particles inhaled is so high and the toxicity so great that it exceeds the defence capacity of the lungs and upper airways, with the added problem that once the lungs are inflamed, they tend to reduce their function. Children, the elderly and people in poor health end up being the main victims. And the effects of this sad and unhealthy culture can be seen in public health centres.

Smoke from fires is serious. In Mato Grosso do Sul, thousands of people from 79 cities make the traditional rush to health centres every winter in search of medical care.

As a recurring problem in most cities in peripheral and/or emerging countries, air pollution has become one of the central public health problems of this new century. According to the World Health Organisation (WHO), up to 3.5 million people die every year from diseases aggravated by smoke, i.e. allergies, sinusitis, flu, bronchitis, embolism, asthma, pneumonia, emphysema, stress and hypertension.

Also, the chemicals in the smoke that are inhaled and enter the bloodstream become another problem for heart patients. According to studies, more than 70 chemical products have already been identified in the smoke resulting from burning waste land.

It was also found that urban smoke is more harmful than rural smoke, because tyres, plastics

and everything else are burnt in the middle of urban vegetation, increasing the toxicity of the air and forcing the public authorities to spend millions on treatment and hospitalisation.
But despite the problems of smoke for human health, urban fires are also disenchanting. Urban voids, when burnt by fire and transformed into ash and soot, end up creating sinister, disheartening landscapes that simply disfigure the aesthetic beauty of cities. And when you want to get to the root of the problem, for some the blame for the fires lies with the dry air, the prolonged drought, in short, with nature; for others, it lies with those who, using the Neolithic method, usually set fire to their plots of land and backyards because of the zero-cost clean-up. It is believed, however, that the real cause lies in cultural heritage, which has already been overcome by various peoples through the valorisation of Environmental Education. Also, when it comes to a situation of urban prophylaxis, it is positive that organised society, in its various segments, points the way to effective social control of this serious problem. A control that takes into account environmental campaigns, legislative improvements, and everything possible to eradicate this evil, if true sustainability begins with the quality of the air we breathe.

DOURADOS FOREST, A RELENTLESS DEVASTATION

"When you hurt the land, you also hurt the children of the land". (Cacique Seattle)

Of all the components of the landscape of Dourados and the region, there is no other that has stood out in the past as intensely and vigorously as the Mata de Dourados, which is the Atlantic Rainforest itself. Coming from the coast, this forest reached the lands of the state of Mato Grosso by spreading through the valleys of the Paraná river basin.
The legendary Atlantic Forest, which originally occupied more than 1.3 million kilometres^2 of national territory, was once a gigantic biome. A biome that has even been included in the National Heritage List for having harboured one of the greatest biodiversities in the world, including both the exuberance of its floral composition and many of the birds, reptiles and mammals of the Brazilian fauna.
However, no more than 7.3 per cent of this forest now remains, given the scale of the relentless devastation that still persists and which makes this biome one of the most endangered in the

world.

In Dourados and the surrounding region, the Dourados Forest, once rich in species such as ipê, cedar, vinhático, balsam, angelim, peroba, faveiro, amendoim, aroeira, jatobá, timbaúva, canafístola, angico, erva-mate, among many others, was surely the most important biogeographical reference in the Paraná river basin and the most expressive vegetation in the territory of the state of Mato Grosso.

But what was once a pleasant, vigorous forest is now just a pale image of the past. Fragmented and confined to tiny areas, it has broken all deforestation records in the state. In Dourados, only 3% of the original cover remains. This is an affront to the Brazilian Forestry Code, which has required the preservation of 20 per cent of forest reserves per property since 1965.

It is essential to open up forest areas for production. The problem lies in the excess. The Dourados Forest, in turn, was untouched until 1943. However, the establishment of the Dourados Agricultural Colony (CAND) alone, according to Teixeira, M.A., covered an area of 300,000 hectares.[11] So far, so good. From 1970 onwards, however, with the expansion of the agro-export model, what we saw was a real environmental disaster, such was the momentum of the devastation.

The scythe and the axe have been replaced by chainsaws and tractors with blades, and fires have increased alarmingly. Hence the escalation of gigantic deforestation, in flagrant disregard for biodiversity, ecological sustainability and the Brazilian Forest Code.

Even so, this forest is still under pressure. Deforestation is relentless and the environmental liabilities are immense. It's hard for the remaining 3% of the forest to escape from the drive for production, despite the appeals in its defence. Local, national and even international appeals, including those from SOS Mata Atlântica.

Nature is certainly not going to let up. In some parts of the country, once rich in vegetation, desertification processes are already appearing. In Mato Grosso do Sul there are already some signs of the problem, considered by environmentalists to be the most worrying on the list of environmental degradation.

11 TEIXEIRA, M.A. Agricultural changes in Mato Grosso do Sul. Doctoral thesis. São Paulo, 1989.

SOIL DEGRADATION IN MATO GROSSO DO SUL

Of the utmost importance, soils, despite their life-supporting status, are a fragile, delicate resource. As pedology teaches us, a few centimetres of soil that can take thousands of years to form can also degrade quickly.

Soil can be considered degraded, according to Embrapa, when it reaches 50 per cent below its productive potential. And this is due to inadequate management practices. Practices whose consequences are not restricted to environmental imbalances, involving resources such as water and biodiversity. It also affects human life in the affected regions. And when it comes to soil degradation, the close relationship of dependence between society and nature becomes much clearer.

The degradation of the soil is equivalent to the disintegration of the laws of survival itself. It is life itself being undermined at its very foundations. Hence the social burden, since the fall in productivity obviously directly affects the supply of jobs and family income, making migration, population emptying and stagnation inevitable, even leaving the municipalities affected with the lowest Human Development Indexes (HDI).

In fact, concern about erosion has become worldwide. And although it is considered to be the main cause of soil degradation, it has only increased, despite the huge losses in productivity. The UN estimates that the total amount of degraded soil on the planet has already reached 2 billion hectares. An area the size of the United States and Canada combined. Therefore, an effective threat to the well-being of humanity. And when it comes down to it, the big dilemma has been livestock farming, which degrades much more than agriculture. For Wagner Pires, Brazil has the largest area of livestock and also the largest area of degraded pasture in the world.[12] Today we have approximately 200 million hectares of pasture and around 70 per cent of it is either degraded or in some stage of degradation. In Mato

Grosso do Sul, according to Armindo Nelvo Kichel, a researcher at EMBRAPA, the lack of correct management and intensive exploitation are factors that could lead the state to lose up to half of its agricultural production capacity. In the whole of Mato Grosso do Sul, of the 18

[12] PIRES, W. ***Degraded pastures? What should I do: reform or recover?*** Circuit Circuit Livestock, 12 August 2016. Available at http:/www.circuitodapecuaria.com.br/

million hectares of pasture, at least 9 million are degraded .[13]

Water erosion, also known as laminar erosion, is the main cause of pasture degradation. In the absence of proper soil management, rainwater that doesn't infiltrate flows downhill, washing away the soil cover and its nutrients with each rainfall. And since it is an almost imperceptible type of erosion, it ends up being the most damaging, because by the time it is noticed, the loss of tonnes of soil has already taken place, degradation has progressed and productivity has fallen.

Unfortunately, the losses are significant, both environmentally and economically. From this perspective, Manoel Cláudio M. Macedo considers that, taking into account only the fattening phase of cattle, the meat productivity of a degraded pasture is around 2 arrobas/ha/year, while in a recovered and well-managed pasture an average of 12 arrobas/ha/year can be achieved.[14]

On many properties, grazing is still extensive, which means that productivity is progressively decreasing. And since soil is a vital resource - practically the only means of producing food - its rate of degradation should not exceed the rate of regeneration. However, in Brazil, millions of tonnes of soil are lost every year, soil that both silts up the springs and ends up in the rivers, whose muddy waters denounce the fact.

It is well known that soil degradation is the last step towards desertification. This is why there is universal concern that this phenomenon is advancing worldwide. According to Evanildo da Silveira, desertification is causing more damage than global warming and natural disasters such as earthquakes and hurricanes. Considered one of the biggest contemporary socio-economic problems, the phenomenon turns arable land into desert at a rate of 12 million hectares a year. If this rate continues the United Nations (UN) estimates that by 2050 the world will have lost an area equivalent to that of Brazil [15][16] .

In Rio Grande do Sul, for example, municipalities such as Alegrete, São Francisco de Assis,

13 KICHEL, A.N. Half of Mato Grosso do Sul's pastures are degraded, says EMBRAPA. Interview with Paula Vitorino, Campo Grande News- Rural. Campo Grande-MS, 05.ag.2011.

14 MACEDO, M.C.M, et alii. Pasture Degradation, Alternatives, Recovery, Renovation and Forms of Mitigation. EMBRAPA, Campo Grande-MS, 1993.

15 PIRES, W. Degraded pastures? What should I do: reform or recover? Circuito da Pecuária, 12.ag.2016. Available at http:/www.circuitodapecuaria.com.br/

15 SILVEIRA, E. The advance of deserts. Rev. Planeta - Meio Ambiente. Ed. 519. Cajamar-SP, 21 January 2016.

Santana do Livramento, Rosário do Sul, Uruguaiana, Quaraí, Santiago and Cacequi have already been affected by the process of arenisation, including the formation of dunes, which, for many, is desertification itself in the Pampas. In Mato Grosso do Sul, the danger of desertification mainly surrounds the regions with sandy soils, basically located in the north-west and south-west regions of the state, where the advance of soil degradation has drawn the attention of the World Convention to Combat Desertification. However, in the state of Mato Grosso do Sul, there are already some interesting measures for recovering degraded pastures, such as the use of contour lines on less sloping land, terraces on steeper land, among various other techniques available from Embrapa, the MS Foundation, the Agronomy Centre of the Federal University of Grande Dourados - UFGD, among other institutions. In addition, there are already owners signing up to programmes launched by the federal and state governments, including subsidised credit lines and grace periods, with a view to combating the problem more effectively, among them PRP-CO (Programme for the Recovery of Degraded Pastures in the Midwest) and ABC (Low Carbon Agriculture). This is a good sign. Perhaps the start of a new, sustainable agricultural model.

AGRICULTURAL EXPANSION, FORESTS AND BIODIVERSITY IN MATO GROSSO DO SUL

Mato Grosso do Sul has become one of the country's main agricultural frontiers. A frontier that is constantly growing, with products such as soya, corn, sugar cane, pine and eucalyptus at the forefront, and the forecast is that both sugar cane and forestry will reach one million hectares each in the coming years. Fortunately, geographically speaking, this expansion is advancing mainly on the millions of hectares in the south of Mato Grosso that are still occupied by degraded pastures. In the case of sugar cane, for example, 87 per cent of its expansion area in recent years has been on this land, according to a Conab survey. This is an auspicious fact because, in the immensity of this degraded space, if, on the one hand, millions of arrobas of meat/year are lost due to the thinning of the herd in the dry season, on the other hand, there are already modern and effective techniques that allow these soils to be recovered, making them productive, through grain crops, sugar cane, forestry, or through pasture-crop / pasture-crop-forestry integration and without the need to deforest to open up new areas.

However, when they exist, it immediately comes to mind that it is a spatially homogeneous degradation, with eroded soils, low yields and that the best thing to do would be to make them productive, even in the form of monocultures or intercropping. However, it should be emphasised that in the immensity of the degraded areas there are thousands of properties and, in this universe, there are certainly hundreds of small forests. And these, despite occupying a minimum of space, if well preserved, will continue to make up the biodiversity of this state.

Strictly speaking, economic development with sustainability has become the great challenge of this new millennium. And one of the United Nations' biggest concerns, highlighted at Eco-92, was precisely the accelerated loss of biodiversity as a natural heritage. These forests are vital assets. As small as they may be, they are floristic formations (trees and/or shrubs), remnants of important biomes such as the Cerrado, the Atlantic Rainforest, whose elimination would end up in environmental losses of potentially valuable species that are duly integrated into the ecological base of fauna and flora and that are there to guarantee the reproduction of a genetic heritage that is thousands of years old for Mato Grosso do Sul.

The Cerrado has already been devastated by half and the once leafy Dourados Forest has already been sacrificed by 97 per cent. A more biodiverse and sustainable future depends on today's actions and one of them is that the forests of Mato Grosso do Sul, in their condition as ecological refuges and remaining genetic banks, albeit tiny, should be spared.

PANTANAL AND HYDROELECTRIC DAMS, EVERYTHING TO FAIL

The Pantanal plain, one of the most remarkable sources of life on the planet, could suffer irreversible damage from the installation of 126 small hydroelectric plants (SHPs) in the arc of the plateaus that surround it. For this reason, it was very prudent that the Federal Court in Coxim, at the request of the Federal Public Prosecutor's Office, granted an injunction on 23 August 2012, halting the issuing of environmental licences for these projects.

The fact is that these licences were granted to each plant specifically, and therefore far removed from the holistic vision that should consider the concomitance of environmental impacts, without neglecting the balance of the situation, something that is fundamental when it comes to the Pantanal biome. A biome rich in biodiversity, an earthly paradise, but vulnerable due to its tropical nature, because once one of its components is altered, the whole can suffer a fatal injury.

It is very likely that the installation of these 126 river dams on the edge of the Pantanal will lead to profound changes in the ecological order of the entire biome.

And you don't even have to have the hundred eyes of Argos to see that such an intervention could have serious environmental impacts. Notwithstanding the importance of the hydroelectric dams, there is no denying the lack of control that could occur in the so-called hydro-regulation of the entire Pantanal basin, since the damming of those rivers could significantly affect the seasonality of the waters, the periodic and natural pulsing of floods and ebbs, something inherent to a millenary hydrological regime.

As a result, it is hard to imagine that the hydrological rhythm, which conditions the biological rhythm of the entire plain, will not be disrupted to the detriment of the intense life of the Pantanal. The problem is serious, since this is not just any river basin. This is the Pantanal basin, a universal heritage whose rivers are not mere freshwater channels, but springs teeming with life, veritable corridors of biodiversity whose destruction will never be restored.

These hydroelectric power stations also run the serious risk of having a shortened lifespan, if we take into account that the damming of rivers and the consequent formation of reservoirs will take place, as we have seen, on the predominantly sandy border plateaus, where deforestation still seems to be far from over, as hundreds of kilometres^2, legal or not, continue to be cleared. This is worrying, as the sedimentation of the 126 reservoirs could happen much faster than imagined.

The natural flow of nutrients, the so-called humus, which, under natural conditions, would cover around two thirds of the immense Pantanal plain as a result of the floods, is extremely important for fertilising pastures, both for cattle and for all the local herbivorous fauna, although the impact on fish must be properly assessed. According to experts, the guinea fowl and the dorado are examples of migratory fish that therefore depend on rivers free of any obstacles. Ladders on the dams would perhaps prevent the end of the piracemas and the consequent damage to the species' reproduction. However, studies indicate that with these ladders, pregnant females in their thousands end up being prevented from reaching the nurseries, i.e. the varjões at the headwaters, to spawn and reproduce their life cycle.

The court ruling states that there is no plausible reason to continue subjecting the Pantanal environment to such considerable risks for another day. Nothing could be more sensible, since what is at stake is the expansion of the electricity sector in an extremely sensitive, complex and vulnerable biome.

South Mato Grosso society must support the Federal Public Prosecutor's Office and the courts for demanding rigorous environmental impact studies in the face of the risks of such projects. In the event that economic interests predominate over socio-environmental ones, it is worth asking whether the planned energy production will in fact generate more savings than the traditional Pantanal economy, which, with more than 200 years of sustainability, has become one of the country's main sources of protein (cattle and fish), and which, together with the growing tourism industry, has only reaffirmed what the Pantanal still is: a universal wonder.

SILTATION IN THE PANTANAL

It's possible that by the thirties of this century the Pantanal plain, a divinely planned piece of the Biosphere elected by UNESCO as a World Biosphere Heritage Site, will be transformed into an immense sandy expanse. And there are already predictions to that effect.

The problem lies in the deforestation of the surrounding arc of plateaus, i.e. the peripantanal. These elevated areas are rapidly losing their original vegetation cover.

Dense vegetation formations of forests and savannahs have been devastated by agriculture, especially large-scale commercial farming. And even in distant municipalities, intersected by rivers that flow into the Pantanal, such as São Gabriel do Oeste, the deforestation of these plateaus has had an indelible impact on the Pantanal plain, through one of the worst environmental evils, siltation.

It is well known that almost all of the problems affecting the interior of the Pantanal are directly related to what happens in the surrounding plateaus. The Pantanal plain, despite being home to the planet's main source of life, is made up of a set of tropical ecosystems, which are therefore fragile and sensitive, and which could even disappear as a result of the environmental pressure coming from the plateaus.

It is worth noting that, under the original conditions, a substantial part of the rain that fell on these plateaus was absorbed and infiltrated into the soil by the vegetation present there, resulting in immense water reserves in these high parts, whose waters were only released slowly through myriad clean water springs, something of the utmost importance especially during periods of drought.

This ensured the hydro-regulation of the entire Pantanal basin, which perpetuated the coming and going of the waters, i.e. the hydrological regime of floods and ebbs throughout the Pantanal.

The Pantanal basin must be understood as a legitimate importer of water from the surrounding plateaus, from where almost all of the rivers that supply it originate and which guarantee the exuberance of its present life. However, these plateaus are still being victimised by deforestation. This is extremely worrying, given the high sensitivity of the sandstone soils on these plateaus and the immense load of sediment that ends up flowing down into the Pantanal basin.

Strictly speaking, siltation has been identified as one of the world's biggest ecological problems. In the Pantanal, as everywhere else, it reduces the depth of rivers, lakes, lagoons, bays and corixos, impacting all forms of local and regional life.

In the Pantanal biome, the damage caused by siltation is already very visible. You can see how much the hydrological rhythm, which conditions the biological rhythm, has been disrupted. Siltation is already occurring in all of the Pantanal's rivers. Including the Paraguay river, the Pantanal's main artery. There, the entire waterway has silting problems, especially during droughts.

And this is a very serious, worrying issue. Areas that used to be flooded only during defined periods (droughts) are now under water all year round. This is the case in the Taquari region, where the river, losing its bed, has permanently flooded an area of around 500,000 hectares, according to data from Embrapa Pantanal. What's worse, around 30 farms are already unusable and totally abandoned.

It's worth remembering that during the rainy season, what flowed into the Pantanal was organic matter, produced gracefully by the vegetation that covered the surrounding plateaus. It was humus that went downstream, not sand. This natural fertiliser, of the highest quality, was carried by the floods and deposited on the meadows, benefiting around two thirds of the great plain. It was the guarantee of beautiful natural pastures and the marvellous variety of

Pantanal life.

Now the Pantanal, this sanctuary of world ecology, is vulnerable. The ghost of deforestation continues to haunt the surrounding plateaus. Of its entire length, very few areas are officially protected. Against this backdrop, it remains to be seen what will happen when the remaining vegetation in these sloping areas has disappeared, or when the entire sandy cover of these same areas has been transported to the interior of the Pantanal basin.

It is known, however, that some landowners are consciously investing in new soil conservation technologies, improved pasture management and no-till farming, among other sustainable actions.

The federal government, through the National Water Agency, has been releasing funds for the recovery of micro-basins and the establishment of seedling nurseries for planting around the springs from the plateaus. There are still around 15 institutions studying and seeking to improve regional hydrodynamics. These are signs of hope, given the extent to which the Pantanal basin is already silting up.

GUARANI AQUIFER, A THREATENED GIANT

The United Nations (UN) has drawn humanity's attention to the problem of the scarcity of water, humanity's most precious commodity. Once considered a renewable, infinite natural resource, it was after the Dublin Conference (1993) that it came to be considered not only a finite resource, but also a scarce and vulnerable one. In this new century, all countries are discussing the scarcity of drinking water, an issue defined by UNESCO as the most serious environmental problem today. This is why it has taken centre stage at events such as the Helsinki Convention, the Doha Conference, the World Water Forum (Japan), the Foz do Iguaçu Charter, the Dublin Conference, Eco-92, etc.

It is universally known that the planet's fresh water reserves are less than 3%, and that more than 2% of these reserves are inaccessible to the 7.2 billion inhabitants. Now, with the vertiginous growth of the world's population, the great challenge is to preserve water sources. According to Bruno Calixto, it is estimated that 4 billion people in the world face water shortages at least one month a year and 500 million people face shortages all year round.[17] Consumption has never increased so much, while water sources are being depleted. Meanwhile, less than 1% of available fresh water is surface water, almost all of whose sources, rivers and lakes, are contaminated, which has led to increasingly intense use of aquifer reserves, not least because they are of good quality.

Despite the little importance that has been given to the subject, millions of people around the world have come to rely on groundwater, drawn from the depths of aquifers. These are confined waters, embedded in sedimentary rocks and which can reach great depths.

In Mato Grosso do Sul we have the privilege of being able to count on a gigantic reservoir, the Guarani Aquifer, one of the largest deposits of fresh drinking water on the planet, which stretches for no less than 213,700 kilometres^2 in the state, and which has been used mainly to supply cities such as Dourados and Três Lagoas.

In its entirety, the Guarani Aquifer, with its 1.2 billion square kilometres, has a water reserve greater than the entire volume of water in the world's rivers and lakes. According to experts, in the event of a possible worsening of the global drinking water crisis, which has already been predicted by the entire world scientific community, this aquifer will be strategic for humanity.

Interestingly, ninety per cent of this immense treasure, comparable to a huge underground water tank, is protected by a gigantic clay cap, formed by basalt spills more than 120 million years ago, from which the reddish soils, considered to be the best agricultural soils in the country, were derived.

The buffer could be a perfect work of nature if it were homogeneous throughout its length, but there are areas of outcropping of the sedimentary rock itself - sandstone. And that's where the danger lies. In these areas of permeable soil, also known as recharge zones, rainwater infiltrates easily and increases the risk of contamination.

It is known that every farmer's dream is to plant on purple soil, which corresponds to the aquifer's clay buffer. However, with the strong expansion of the agricultural frontier, the sandy soils mentioned above are also being occupied by huge plantations of soya, corn and

17 CALIXTO, Bruno. Scientists estimate how many people live with a lack of water in the world: 4 billion. Época Magazine, Rio de Janeiro, 01.nov.2016.

sugar cane.

The thing is, these porous soils make it easier for rainwater, in contact with pesticides and chemical fertilisers, to contaminate both the water table and the aquifer itself. Not that clay soils (which are more impermeable) are immune to the problem. Basalt (the cap's parent rock) is known to contain cracks (diaclases) through which poisoned rainwater can penetrate.

And this, combined with the intense application of agrotoxins and chemical fertilisers on the giant (clay buffer), with a view to the production of "commodities" (soya, maize, sugar cane...), make up a serious, risky binomial in which there are few warning signs.

It should be emphasised that, according to the National Council for Food and Nutritional Security, Brazil leads the world in the consumption of pesticides and, according to the National Health Surveillance Agency (Anvisa), the sector has grown by 190% in the last 10 years[18] and, since the soils of the Guarani Aquifer are the most arable in the country, it is likely that they are also the ones that receive the largest volumes of pesticides and present levels of contamination at least close to the risk limit for human health.

In fact, the Guarani Aquifer has never been so threatened and the warning cry must be resounding, because underground aquifers, once contaminated, are contaminated forever. This is unfortunate, as it is a strategic reserve of drinking water that could serve humanity itself. Although sustainable management projects for this aquifer are underway, so far there is no known monitoring system to protect it. It is known that Embrapa has been studying the issue and, rightly, recommending the reorganisation of agricultural production in these areas. In the meantime, all scientific communities are practically emphasising the need to change the current production model. And no wonder. Imagine the damage that some of these agrochemicals can cause, especially those banned in their countries of origin, when released into the recharge zones of this aquifer. The damage could be irreversible.

In Mato Grosso do Sul, water resources have been thought of almost solely in terms of surface water. It's time to think about groundwater. Given its vulnerability, the Guarani Aquifer in this unit of the federation needs a permanent monitoring network, within a multi-institutional strategic plan, which will generate research and lead to solutions, while there is still time.

18 National Council for Food and Nutrition Security. Studies show a link between pesticide consumption and cancer. Source: Ascom/Cosea, Brasília-DF, 08.Apr.2016.

THE CERRADO'S WATER TANK IS RUNNING DRY

With the focus of environmental concern always on the Amazon, little is said about the devastation of the Cerrado. The world's richest savannah in terms of biodiversity, the Cerrado, which is also the country's second largest biome, is under serious threat. And it's hard to believe that half of this beautiful floral formation has already been destroyed by deforestation. In globalised times, this immense biome, with its more than two million km^2 and gentle topography - plateaus and chapadões - ended up becoming a perfect setting for the advance of the so-called agricultural frontier on a large scale. The result is that thousands of native species have been replaced by monocultures of cotton, sugar cane, eucalyptus, soya, maize and, above all, livestock.

Based on biotechnology and agro-export, the Cerrado biome has already become Brazil's main grain, meat and milk production area. Nothing better from the point of view of food production, but a paradox in the face of development with environmental sustainability. After all, it is worrying to realise that the cerrado, which is present in 12 federal units, has been the Brazilian biome most heavily affected by devastation.

Thus, for Alexandre Mansur, the devastation of the Cerrado is greater than that of the Amazon... In 2014 alone, the Brazilian Cerrado lost 6,000 square kilometres. In a decade, the savannas and forests of the Cerrado have lost the equivalent of an area larger than the state of Rio de Janeiro. And this devastation has several consequences. The first is the destruction of biological diversity. The second is the jeopardisation of some of the source areas of the country's main hydrographic areas.[19]

Nobody has the right to be against food production, but the preservation of strategic areas of this biome would be fundamental in many ways, and it is important to emphasise it as a water reserve for the country.

This is a serious problem. The impacts of the devastation of the Cerrado on the national hydrological system have been irreversible.

With its thousands of springs, the Cerrado biome was once considered the "cradle of water"

19 MANSUR, Alexandre. The Cerrado has lost a Rio de Janeiro in ten years. Find out the consequences. Blog Planeta Época. Rio de Janeiro, 01.nov.2016. Available at http://epoca.globo./ciencia- e-meio-ambiente/blog-do-planeta/

due to its crucial role in supplying Brazil's hydrographic basins. However, with deforestation, its springs, impacted by silting, are no longer flowing. With this in mind, the electronic magazine Ecodebate, when addressing the issue, says: A nursery of springs, with its high topography and strategically located in the central *Brazilian* plateau, *the Cerrado functions like a gigantic water tank. Eight of the* country's *twelve* river basins are irrigated by the biome. 18 20

It's sad, but true, that the Cerrado, this magnificent biological heritage of the nation, is already on the list of the most threatened biomes on the planet. And no wonder. But the cerrado biome cannot end. The damage would be too great, even for agribusiness itself. And it's no use simply reducing the rate of deforestation. It has to be eliminated. In order to recharge the soil's water reserves, it is essential that production areas are harmonised with strategic areas of vegetation.

It is also essential to expand the preservation units (UPA), because the ones already in place so far are nothing more than islets in the immensity of the devastated areas.

SAVANNAH AND CHARCOAL PRODUCTION IN MATO GROSSO DO SUL

The 21st century is progressing and, in the middle of its second decade, the Cerrado biome continues to be relentlessly devastated, despite its inestimable value as the source of a large part of the country's water springs. Its greatest losses are attributable to the state of Mato Grosso do Sul. In that state, a huge area was already lost in 2013. In that year, according to the Ministry of the Environment (MMA), 68 per cent of its original cover had already been cut down in order to make way for agriculture, and this activity has continued, covering huge areas, especially with commodities. And in the wake of deforestation, the presence of charcoal, as a by-product, and at the end of the black diamond production chain, thousands of charcoal factories spread across the state, which, extrapolating geographical limits, reached the Pantanal, a unique feat.

This is the type of charcoal often consumed in homes, bakeries, steakhouses and pizzerias,

20 With almost 20,000 springs, the devastation of Cerrado basins impacts the lives of 88.6 million people. Ecodebate - News, by Redação. 25.Apr.2011

and it has attracted a lot of interest from steel mills, whose importance lies in the fact that, together with iron ore, they produce pig iron, the raw material for steel, a product that is widely used in industry.

It so happens that, at the mercy of the interests of farming and cattle raising, as well as steel mills, a substantial part of this biome - the Cerrado - has been the victim of totally illegal deforestation, true environmental crimes, leaving ecological footprints with serious consequences for biodiversity, And what's worse, without improving the lives of thousands of workers at the mouth of the furnace - the charcoal burners - and their families, who continue to make up the sad scenario of human and environmental degradation, with reports of child labour and labour analogous to slavery all over the world.

Among the many reports on the devastation of the Cerrado, here's one by Rodrigo Vargas in 2010: The production of charcoal for the steel industry has caused around 270,000 hectares of native forest in the Pantanal of Mato Grosso do Sul to disappear in the last three years and the advance of charcoal factories over native forests, whether legal or not, is a serious threat to the survival of the Pantanal.[21]

According to IBAMA, in 2005, amazingly, there were 5,000 charcoal factories operating in the state of Mato Grosso do Sul. Of these, only 468 were legalised. And how can we believe that the situation has improved since then, when the public institutions responsible for inspection are routinely short of staff, vehicles and budget?

On the other hand, Mato Grosso do Sul impresses with its contemporary planted forests, mainly eucalyptus. This can be seen in the report by journalist Caterine Sturza: *"Mato Grosso do Sul currently has 920* million hectares of eucalyptus plantations, with Reflore-MS and Famasul projecting that the state will reach 1 million hectares by the end of the year...and that it *currently has the world's leading pulp companies, Fibria and Eldorado* [2]. [22]

However, despite the countless benefits they bring to Mato Grosso do Sul, it is worth remembering that the companies establishing these forests initially signalled their intention to produce enough biomass both to feed pulp production and to supply the steel mills, thus

21 VARGAS, R. In Mato Grosso do Sul, 270,000 hectares turned into coal in three years. Folha de São Paulo - Science. São Paulo, 05.Apr.2010.

21. STURZA, C. MS will become the country's largest pulp producer from 2017. Midiamax, Campo Grande- MS, 18.ag.2016.

putting an end to the devastation of the Cerrado and other biomes in the state.

But unfortunately, that's not what happened. Despite the fact that the projections for 2017 show Mato Grosso do Sul as the national leader in pulp production, little of the charcoal from the planted forests (forest massifs) has been used by the steel mills. In fact, almost all of the eucalyptus and pine plantations have been used to produce pulp and paper, not charcoal. So, despite the many commitments made to end the devastation of the biomes of southern Mato Grosso, the steel mills continue to receive the *sacred charcoal*, often from clandestine, illegal deforestation. It's unbelievable, but such carelessness is still going on. By simply consulting the media, current headlines (2017) show the recurrence of the problem, albeit on a one-off basis, given the limitations, as we have seen, of inspection. Here's one example: Police close charcoal factory in Mato Grosso do Sul and seize 161 metres^3 of firewood and charcoal.[20 * 23]

So while the illegal sale of charcoal to steel mills continues, the disfigured Cerrado remains the biggest victim of devastation in Brazil. Incredibly, according to the renowned South Mato Grosso NGO Ecoa (Ecology and Action), this biome has a deforestation rate 2.5 times higher than the Amazon.[24] And yet, a bill banning the production, transport and use of charcoal from native forests in the state of Mato Grosso do Sul, authored by state representative and environmentalist Amarildo Cruz, was vetoed, despite unanimous approval in plenary.

In the immensity of the territory of Mato Grosso do Sul, there is no shortage of space for the advance of agriculture. There are around 9 million hectares of degraded pastures, with very low productivity, which, although some have already taken the initiative to reuse them in a sustainable way, are still blatantly encroaching on the remaining biomes. Charcoal factories don't just appear. If rural landowners weren't still willing to deforest and steel mills were eager for the black diamond, the charcoal production chain would obviously no longer exist and the Cerrado biome wouldn't be as disfigured in its physiognomy as it is in Mato Grosso do Sul.

23- Police close charcoal factory in MS and seize 181 cubic metres of firewood and charcoal. Globo.com - G1 MS, Campo Grande-MS, 08.jan.2017.

24 Rescuing the Cerrado from Extinction. Ecoa, Campo Grande MS, 07.Dec.2016 - Available at riosvivos.org.br/notícias- ecoa/resgatar-o-cerrado-da-extincao

RIPARIAN FORESTS CRY OUT FOR HELP IN MATO GROSSO DO SUL

It became apparent that on the banks of the Dourados, Brilhante, Vacaria, Ivinhema, Amambai, Iguatemi, Miranda, Pardo, Paraná and Paraguay rivers, among others, the riparian forests, with the exception of a few properties where rural producers consciously keep them intact, were being devastated, losing their status as primary forests to secondary forests.

Although these formations, also known as gallery or riparian forests, are the main protectors of water sources and are therefore considered permanent preservation areas in the Brazilian Forest Code , despite the multiple environmental functions they fulfil, they have unfortunately suffered an intense process of deforestation both in Mato Grosso do Sul and in the majority of Brazilian states.

Deforestation that, in many cases, has already fallen far short of the widths required by law, including unprotected riverbanks where rivers run in full sun. Strictly speaking, leafy trees of up to 30 metres have been felled since the beginning of territorial occupation, and today there are situations where crops and/or pastures reach the banks of rivers, eliminating precious stretches of these forests.

Impacts such as cattle trampling, invasive plants, sand extraction, drinking fountains - except for the well-designed ones - and fires, including in park areas, have been directly affecting this very important plant formation.

As a result, rivers, lakes and streams have not been spared from environmental degradation. There are floods, landslides, silting up, increasingly murky, shallow and muddy waters, which add to the disappearing springs. Biological corridors bordering rivers are being destroyed and transformed into islands where fauna and flora are confined. They are deforested stretches that freely expose bodies of water to agrochemical products, including frequent fish kills.

Certainly, the global ecological crisis, so present in Mato Grosso do Sul, is best reflected in the waters of the rivers. However, the riparian forests of this state, even though they are in a varied state of degradation, are still subject to recovery, for the sake of the quality of these waters.

What we mustn't do is put it off, otherwise what has been happening around the world will happen to the water sources in the south of Mato Grosso. Legendary rivers such as the Colorado in the United States and the Yellow in China are already drying up at certain times before they even reach the sea, deprived of the protection of riparian forests.

In the south of Mato Grosso, the Taquari River, which has already made around 30 rural properties unviable, is a reminder of the need for society to move in harmony with nature, seeking production but without giving up vital preservation spaces.

Fortunately, riparian forests - the link between terrestrial and aquatic systems, and therefore essential for sustainable rural development - are already starting to feature on the water management agenda in several countries, including under the guidance of the UN. In Brazil, good initiatives have already reached the states of São Paulo and Paraná.

In these times of environmental awareness, technology and sustainability need to point towards a new development model, where good practices for preserving and restoring natural resources can take place.

CHAPTER 4

PARANÁ RIVER, HYDROELECTRIC PLANTS AND INCREASED RAINFALL

A developing state like Mato Grosso do Sul cannot do without a more in-depth knowledge of its resources in order to make better use of its potential. It is common knowledge that the success or failure of nations depends directly on the way in which their natural resources are utilised.

From this perspective, there is the possibility that a strip of land in the south of Mato Grosso bordering the Paraná River, covering approximately 2.5 million hectares and stretching from Três Lagoas to Mundo Novo, is benefiting from an increase in soil humidity due to the evaporation that is rising from the large reservoirs formed by the dams of the Jupiá, Ilha Solteira, Porto Primavera and Itaipu hydroelectric power stations, which has increased the rainfall around these reservoirs.

This strip, some 50 kilometres wide and 500 kilometres long, bordering these gigantic water reservoirs on the Paraná River, whose mirror can be up to 12 kilometres wide, is actually bordering one of the largest dammed water bodies on the entire surface of the earth. It is therefore very likely that it is receiving an increase in precipitation.

In fact, the presence of large areas of evaporation could be increasing the so-called convective rainfall on the banks of the Paranazão. Just imagine the volume of water vapour that rises from this river in its 500 kilometres of impoundment.

A comparative analysis of how much it rained before and after the installation of these four hydroelectric dams would make it possible to know the behaviour of the rainfall on this strip of land. Although some of this water may enter the upper atmospheric circulation and precipitate over long distances, another part may be precipitating over this 2.5 million hectare strip of land through the so-called edge rainfall effect.

If this is confirmed through research, we would have a huge geographical area benefiting from natural and totally free irrigation. And since water is the most important agricultural input, it would be interesting for this space - if the increase in rainfall is confirmed - to be replanned, in view of a much greater gain in productivity, if it were used more intensively, especially by family farming, using the good practices of agroecology, in order to avoid, for example, contamination of those immense water reservoirs.

Agroecology, which has been recognised worldwide as the way of farming that is best integrated with the laws of nature and life, basically using smallholdings and diversifying crops, has led to the emergence of orchards, woodlands, gardens, vegetable gardens and windbreaks, which end up reaffirming the principle of environmental biodiversity and sustainability.

This possible change in the economic vector, with a view to making better use of the moisture in the soil (2.5 million hectares), which would basically shift from extensive and semi-extensive livestock farming to family farming, would take place through the peaceful acquisition of these lands through the National Institute for Colonisation and Agrarian Reform (INCRA).

And that it should be through a model of technological agrarian reform, where, for example, for every group of 100 plots, ten would be allocated to technicians, both trained and landless, who would pass on the technologies to the rest.

After all, in Mato Grosso do Sul, good rural schools and federal institutes, as well as several universities, have been training these technicians. It should also be emphasised that, in addition to the possible increase in water, the region in question already has good logistics for transporting crops, given the presence of an extensive asphalt network and the Paraná-Tietê waterway. Throughout history, we know that not a few societies have fallen into decline due to the misuse of their natural resources. Unemployment, low quality of life and population depletion are some of the bills paid. And in this immense strip of land, which could be very well served by natural and free irrigation, what we see is that it still suffers from one of the lowest population densities in Mato Grosso do Sul and in the country, despite its neighbourhood with developed states such as São Paulo and Paraná.

Now, while it is true that Mato Grosso do Sul needs to develop its productive potential by making better use of its natural resources, it cannot live with the immensity of degraded pastures in the hands of absentees, for example, which are so common in that precious strip of land.

In fact, for a new dynamic in food production, a new boost in the generation of jobs and income, a new increase in municipal and state public revenues, as well as a rise in the demographic density of that area, regional re-planning is necessary. After all, Mato Grosso do Sul has respectable institutions dedicated to this field of study.

SUSTAINABILITY, THE WORD OF THE CENTURY

A modern, socially advanced society cannot be envisaged without sustainability. From the moment humanity realised that natural resources are finite and that the environmental risk of economic development needs to be taken seriously, sustainability became the word of the century, and sustainable development, in turn, became the basic concept for building a new development model. A model that satisfies today's needs without jeopardising the lives of future generations.

In the face of the environmental pressure, the mentality that has been established worldwide, especially since the famous Agenda 21, is no longer based on development at any cost (developmentalism), but rather with quality, respecting, of course, the limits of the environment. It's unfortunate, but development is still very much confused with economic growth. And it's no coincidence that degrading activities continue to be encouraged, without taking into account the depletion of natural resources, on which the links of nature, life and human existence itself depend. We can no longer accept the concept of development evaluated solely in the light of the expansion of material wealth, of economic growth, because just as important as the generation of wealth is the preservation of the environment and the quality of life of the population.

It is unbelievable that, today, the main indicator of countries' development is the Gross Domestic Product (GDP), which does not measure sustainability, to the detriment of the Inclusive Wealth Index (IRI) proposed by the UN.

In Mato Grosso do Sul, although we don't yet have a specific IRI, it can be seen that many actions are still guided by the anti-ecological model, with unbridled exploitation of natural resources. As a result, the processes of degradation are very present there, with fires (the municipality of Corumbá ranks first in the country), illegal deforestation (including riparian forests), soil erosion, hypermonoculture, pasture degradation, increasing volumes of pesticides and chemical fertilisers, silting up and polluting rivers, obstructing springs and the constant loss of biodiversity.

On the other hand, there's no denying that, alongside these imbalances, some good practices have been used, such as the application of contour lines, terracing, straw planting, crop-livestock integration, biotechnology and agroecology, among others, which have in fact resulted in increased production, while protecting and enhancing the environment, as befits sustainable development.

The sustainable model, it should be emphasised, does not condemn economic development. On the contrary, it promotes it, focussing on the environment as its main ally.

BIODIVERSITY AND HUMAN HEALTH

The importance of biodiversity for human health has already become an important topic at United Nations international conferences, recognising it as a precious source of medicines. The World Wildlife Fund has also drawn attention to the need to curb deforestation and preserve biodiversity in the name of human health.

Not surprisingly, almost all of the medicines produced by traditional medicine involve the use of plant extracts, therefore remedies that originate in nature, and biotechnology has been searching for new means of healing based on new chemical components or active principles from biodiversity products, with the discovery of new antibiotics, antiviral agents and vaccines.

The tendency to valorise natural products and the consequent substitution of synthetic products by plant-based products has been increasing. According to Lima, J., human beings have always used plants to relieve pain and suffering. The World Health Organisation (WHO) estimates that 80% of the world's population regularly relies on medicinal plants for their primary health care.[25]

And it's crystal clear how far phytotherapics and homeopathics have taken medicine by providing solutions to many health problems.

Strictly speaking, modern medicine owes a lot to the properties found in medicinal plants, revealed on the basis of traditional knowledge. Today, the world's pharmaceutical industries

25 LIMA, J. Plantas e Dor. Contribution to the Anthropological Study of Pain Treatment. Almada, University of Coimbra, 2010.

have research programmes in the field of natural products and, as we know, important drugs such as morphine, aspirin and many others have been synthesised from plants. However, the world is in danger of missing out on new treatments for cancer, osteoporosis and other diseases if it doesn't move quickly in the direction of biodiversity.

The loss of biodiversity is serious. Several medicines that could cure many ailments are no longer found. Plants that have not even been studied for their therapeutic efficacy have already been wiped off the map and the famous red list of threatened species, recently presented by the International Union for Conservation of Nature, is frightening. It now stands at 16,000 species.

Drawing inspiration from Ulrich Beck - *think globally and act locally* - it is essential to remember how much the territory of Mato Grosso do Sul also *groans in labour pains*, because the speed with which this storehouse of biodiversity is being reduced by burning and deforestation is incredible, without being able to predict the loss of strategic species for the future, in the case of medicine.

We humans, while being directly dependent on this divinely planned marvel - biodiversity - are also the only ones capable of destroying it, as has been happening. However, we are comforted to know that we are also the only ones who can reverse this trend. The choice is ours!

BIODIVERSITY IN MATO GROSSO DO SUL

In this Third Millennium, while Mato Grosso do Sul has managed to establish itself as one of the country's main agricultural frontiers, it has also experienced a clear and marked loss of its biodiversity because, in addition to the expansion of grain production areas, such as the soya/maize binomial, sugar and alcohol mills and cellulose factories have also demanded ever larger areas for sugar cane and eucalyptus, thus expanding the hyper-monoculture areas.

As a result, biomes in the south of Mato Grosso, as complex and diverse as the Cerrado and the tropical forests, which are extremely rich in plants and animals, are being replaced very quickly. This has led to the formation of homogenous, simplified landscapes, which obviously go against ecological precepts.

At this point, although everyone knows how important biodiversity is, the losses are more evident than ever. Plant and animal habitats are rapidly disappearing. Biomes everywhere are being reduced and, with them, biodiversity is in full decline. In the face of this, we believe there is a consensus - that everyone should recognise biodiversity as an irreplaceable natural heritage. A vital heritage for humanity. But it is also severely threatened, with species disappearing on an ever shorter time scale. It's hard to imagine that more than 100 species are becoming extinct from the planet every day. Some say as many as 150.

Whether due to anthropogenic action or natural processes, it is certain that the rate of biodiversity loss is far greater now than at any other time. And this is something serious that needs to be evaluated in every corner of the world, especially in monoculture areas like ours, because the race is against time.

And it's not a question of choosing which species to preserve. Each component of the Biosphere, being the fruit of millions of years of organic evolution, must have its value: predators regulate the expansion of prey, pollinators increase fruit and grain production, photosynthesising plants reduce carbon dioxide in the atmosphere, and what about biodiversity as the source of most medicines?

From this point of view, it should also be seen that most of the species that go extinct every day were never even known to science, and what is being lost in some cases are real gene banks that will certainly be sorely missed.

Rural production, a mission that is recognised as noble, sacred and essential, no matter how much it needs space for planting, should not give up on preservation. Preserve, even the minimum. Preserve at least 20 per cent of legal reserves per rural unit.

Facts such as the felling of 97 per cent of the Dourados Forest require a lot of reflection and reparation. Let's remember that Mato Grosso do Sul was once one of the most biodiverse areas in the world. And that fascination and enchantment with Mato Grosso do Sul's natural landscape, so well represented by its tropical forests, savannahs, wetlands and also by the grace and magic of its fauna, is fading into the background. And it's not even a question of returning to the past.

Of course, the wheel of production cannot stop, but the accelerated expansion of monocultures, even if it turns Mato Grosso do Sul into a veritable agricultural granary, should never be disconnected from nature. Preserving it is essential, even if only to a minimum, because surely not even the latest technology, such as precision technology, will be able to compensate for the loss of this heritage - the heritage of biodiversity, the heritage of humanity!

ECOLOGICAL CORRIDORS, A CHANCE FOR LIFE

In recent years, with the new expansion of the agricultural frontier, Mato Grosso do Sul's biomes, such as tropical forests, savannahs and other plant formations, are being visibly reduced to fragments, i.e. small islands of biodiversity, due to the new wave of deforestation. Fragments, which are quite scattered and surrounded by monocultivated areas, have had a marked and negative impact on the remaining fauna and flora in these areas.

Geographically, wild plants and animals from the rich biodiversity of Mato Grosso do Sul have been suffering the tragedy of spatial confinement, such is the degree of reduction and isolation of their habitats, whose preservation is becoming increasingly difficult.

These fragments, if they continue as they are, have no way of ensuring what everyone wants - sustainability. And despite their biological importance (because they're what's left in the immensity of the monocultivated spaces), most of them are abandoned, in the form of secondary vegetation, because what was good and best, such as leafy trees, has already been extracted.

In fact, the degradation of these fragments has been a constant, as most of them have been disturbed by clandestine deforestation, invasive plants, fires, pesticides and a lack of water for the fauna. Also, the distance between these fragments, making connectivity between fauna and flora difficult, can only hinder the spread of pollen and seeds, as well as gene flow between species. And because of the distance between the fragments, as well as the isolation imposed on the fauna, the problem of inbreeding (mating between individuals of the same species) arises, which weakens the health and accelerates the process of extinction of these animals.

Conserving biodiversity in intensively cultivated areas has been a major challenge for the interconnection between forest fragments.

Fortunately, this is becoming perfectly possible through so-called ecological corridors. Something that is being considered worldwide as a modern strategy. A strategy that only tends to improve local and regional biodiversity, as well as the Biosphere as a whole, by establishing the essential network of relationships between these fragments and their respective species. What's more, if well preserved and connected, they can provide good ecosystem services to properties: helping to control pests and diseases, retaining water in the soil, providing seeds for reforestation, improving the microclimate, pollinating crops and orchards, and even supporting biotechnology, the science of today's century.

Despite its woes, the importance of biodiversity is something that is being increasingly realised. After all, everyone wants to preserve the natural processes that support life. Even more so when it comes to the broad spectrum of life that so well characterises biodiversity in the south of Mato Grosso.

GROSS DOMESTIC PRODUCT AND SUSTAINABLE DEVELOPMENT

It has long been a question of whether the Gross Domestic Product (GDP) measures anything meaningful, because it is an index that, although used worldwide, does not show whether things are getting better or worse for millions of people.

By only measuring quantities produced, this indicator has only increased the obsession with economic growth - growth at any cost. For many, it is an index that is neither humane nor social, given its indifference to poverty and inequality, since its objective is not to measure sustainable development, which, in addition to economic development, also takes into account socio-cultural and environmental components.

Economic growth has long been confused with social development. Growth, although necessary, is not enough, as it does not guarantee development, which in turn is a process of economic, social, cultural and environmental transformation, which implies an improvement in the well-being of society as a whole. A rise in everyone's standard of living.

That's why an indicator like GDP, which focuses only on the economy, loses its raison d'être, as it doesn't respond to the ultimate goal, which is development in the true sense of the word - socially inclusive and environmentally sustainable, as explained so well at United Nations conferences such as Eco 92 and Rio+20, among others.

However, despite the avid interest that still persists in GDP, it is important to know that in many countries, for millions of excluded people, economic growth has only meant more hunger and misery, proving that high GDP levels do nothing to guarantee sustainable development.

Incidentally, the two giants of Asia, China and India, despite having enviable GDP figures in recent years (8% to 10% per year), have not improved the lives of the vast majority of the population, something that is clear from the statement by the president of the Ethos Institute for Social Responsibility, Ricardo Young, echoed by journalist Luciano Martins Costa: The debate about growth rates is pure stupidity. The 9 per cent annual growth rate of the Chinese economy, which the media praise, corresponds to a disaster of equally considerable proportions in terms of sustainability.[26]

In India, when Daia Florios, based on studies by the United Nations Children's Fund (UNICEF), informs us that 600 million people defecate in the streets of cities[27] , the debate doesn't even need to be deepened.

In Brazil, too, in the days of the Brazilian miracle, the country achieved record GDP levels (over 10 per cent a year). The famous Delphinian cake grew without reducing the country's social and economic inequality.

GDP has been widely criticised, including by most Nobel Prize-winning economists, precisely because it ignores social exclusion. And as for the environment, they preach abandoning the myth of infinite economic growth, inspired by GDP, in the face of the exhaustion of the planet's natural resources. For them, it's possible to prosper without destroying.

By instituting biodiversity policies, Costa Rica has managed to double its forest area and at the same time triple its per capita income, demonstrating that it is possible to turn sustainability into a rule of life, even with an unimpressive GDP.

Fortunately, not a few scholars are discussing replacing GDP with a new indicator that takes sustainable development into account and, alongside economic variables, also incorporates socio-environmental ones. The UN General Assembly itself has already passed a resolution describing GDP as an indicator that does not reflect the well-being of a country's people. Even

26 Costa, Luciano Martins. Observatório da Imprensa. Ed. 410, 05/Dec/2016. Available at URL of article: www.observatoriodaprensa.com.br.

27 Florios, Daia. I will do good to the Earth. Greenme.com.br. 09/Nov./2015. Available at https://www.facebook.com.

so, the trap has not been broken.

GDP aside, Brazil, like few other countries, has made interesting progress in recent years in its qualitative performance, becoming, for many analysts, a world benchmark. Millions of Brazilians have been socially included in the last decade alone.

It follows, then, that a high GDP is not enough to make Brazil a world economic power. We already have that! What matters to the vast majority of the Brazilian people is that we are a modern nation with a humanist and happy vocation.

A sustainable nation!

GREEN ECONOMY AND SOLIDARITY ECONOMY AT RIO+20

At the Rio+20 World Conference, one of the key themes was the Green Economy, defined by the United Nations as an economy that results in improved human well-being and social equality, while significantly reducing environmental risks and ecological scarcity. There was no shortage of arguments in its defence. For some, it gains importance because it takes into account the concepts of sustainable development, concepts that will encourage companies to adopt fair, inclusive social practices that demand the eradication of poverty. Environmentally responsible, correct practices that reduce ecological footprints, tackle the vast contamination of ecosystems, the reduction of biodiversity, climate change and are capable of maintaining economic growth.

From this perspective, the transition to a green economy would result in higher global GDP growth rates and higher employment levels in the medium and long term. And at the centre of the discussions is the question of how to reverse the planetary trends of pollution , global warming, water pollution (rivers, lakes, oceans), reductions in the ozone layer, polar ice caps, glaciers, biodiversity and natural resources.

The agenda includes tasks such as reducing fossil fuels (petrol, diesel and coal), progressively switching to clean and renewable energy sources (solar, wind, geothermal, tidal and organic), proper waste treatment (with efficient recycling systems), investments in forestry, fish farming, responsible tourism, basic sanitation, recovery of degraded areas, among others.

At this International Conference, while some renowned speakers believed that the Green Economy could be fitted into the concept of sustainability, others, such as environmentalists and leaders of social movements - representatives of the People's Summit - ended up rejecting this idea, citing a crisis of confidence.

For them, the greening of the economy will not bring about substantive changes in the economic model, which, because it seeks, as it always has, unlimited and infinite growth, has no way of respecting the physical limits of nature, limits that cannot be exceeded, otherwise humanity itself will be put at risk. Finally, they criticised what they called the transition to a Green Economy, defended by the representatives of the rich countries, claiming that the green transition will not be able to overcome the great socio-environmental crises that are currently taking place, because it does not question the foundations of the current economy, because it does not touch on crucial issues such as productivism and consumerism - the roots of predatory growth and ecological scarcity.

On every continent and in every country, there are growing economic and social initiatives in favour of an alternative economic model that organises production, distribution and consumption based on social technologies, creating a parallel universe of products that satisfy people's basic needs based on popular knowledge, in a simple, low-cost way, generating thousands of jobs and income in the countryside and in cities. This is the Solidarity Economy, centred on agro-ecology, micro-enterprise, incubators, micro-credit, social currencies, points of culture, community radio stations, the exchange of products and services, citizen participation in local development and a strict focus on environmental preservation.

For Hélio Mattar, we have made progress towards a greener economy. The problem is that, in essence, nothing has changed.[28] Paul Singer, a renowned USP professor, has said on several occasions: the Solidarity Economy is the best developed model so far. In Brazil and around the world, it is growing all the time.

SOLAR ENERGY, EXCEEDING EXPECTATIONS

Faced with a humanity stunned by the greenhouse effect, global warming, melting glaciers, rising sea levels and increasingly scarce natural resources, it's no wonder that scientists and environmentalists around the world are welcoming solar energy. And no less than the United

28 MATTAR, H. Vitae Civilis discusses transition to a green economy. Conference Ethos,11.06.2012. Available at siteuniethos.org.br.ci2012/?=888

Nations International Conferences on the Environment have been signalling, while the time is still right - the need to overcome the era of fossil fuels (coal and oil) and nuclear power for the era of clean energy sources, with a view to a sustainable future.

And in this race against time, the transformation of solar energy (captured through photovoltaic solar panels) into electricity has been pointed out as one of the most promising paths in the field of energy generation and security, given the prominence it has been gaining in the world media, given the speed with which it has been occurring.

Several countries have been researching and embracing solar energy, both because it is clean, inexhaustible and comes from the sky for free, and because it is practical, as well as being 100 per cent ecological, it is easy to install and maintain. It doesn't involve the use of turbines, motors or other mechanical systems, and it takes advantage of already installed transmission lines.

This is no passing wave. The expansion of solar energy has become irreversible. It is set to become the most important source of energy in the future. To give you an idea, according to the International Energy Agency, solar energy generation capacity has been growing at around 40 per cent a year since 2000 and, at this rate, it will certainly become an extremely important source in the composition of the Brazilian and world energy matrix.

The solar energy sector is certainly undergoing an impressive global expansion, given the speed with which it is doing so. The production of solar panels (photovoltaic panels), for example, has exceeded expectations.

Cars and even transoceanic vessels are already powered by this energy. The first photovoltaic plane, Impulse I, has already made a 6,000 kilometre intercontinental flight between Switzerland and Morocco, half of which at night, and Impulse II has already flown around the world. Every two years, a solar car race takes place in Australia. In Brazil, the Maracanã, Mineirão, Mané Garrincha and Itaquerão stadiums, among others, have been transformed into solar stadiums for the World Cup.

Mega and giga projects based on this energy to supply buildings, commercial, industrial and residential spaces, housing estates and offices are already underway around the world. China, the United States, Portugal, India, Spain, South Africa and Australia are the countries leading this great race.

Los Angeles already has a project underway to supply one million people, and is expected to sequester carbon equivalent to the emissions of three million cars. The solar plant in Mojave, also in California, aims to be the largest in the world. The one in Arizona supplies 70,000 families.

The United States, for its part, has been preparing its southeastern region to become the largest solar energy hub on the planet, signalling that it is on a trade war footing with China in order to take over the world leadership in the sector, which is certainly the first green, carbon-sequestering war - the first healthy and welcome war in history.

In Germany, more than 400,000 new jobs have been created recently through the expansion of photovoltaic solar energy, which has certainly freed Germany from the crisis that the eurozone countries are going through.

In Brazil, a tropical country par excellence, it is already said that we have a veritable pre-salt of solar energy, as well as one of the largest reserves of quartz in the world, from which silicon is extracted - the raw material used to make solar panels. And as we wake up to this race, small and medium-sized plants have already sprouted up all over the country.

And when you realise that several Brazilian companies and institutions, such as universities, are researching and/or investing in the sector, and that important events on the subject are starting to take place, such as the International Solar Energy Technology Fair, you can be sure that it is a priority on the Brazilian agenda, in view of self-sufficiency in clean energy, which is possible by combining hydraulic and solar sources.

It should also be emphasised that since December 2012, according to Resolution 482 of the National Electricity Agency (ANEEL), anyone who generates photovoltaic solar energy (homes, businesses, industries, churches, grain dryers, public buildings), in both urban and rural areas, will be able to receive their bill at the end of the month paying little or nothing for consumption. And the best part: you can get credit from the utility company if you produce more than you consume. In this case, the electricity clock turns backwards, producing credits. In practice, it's the consumer becoming a generator of clean energy, with the right to compensation. Brazil does not yet produce photovoltaic panels on a commercial scale. That's why their cost, although still somewhat expensive by Brazilian standards, has been falling steeply.

The market, induced by the explosion in manufacturing, mainly Chinese, has already become competitive. Jessica Shankleman & Christopher Martin reveal that since 2009 solar energy prices have fallen by 62 per cent, with all segments of the supply chain cutting costs.[29] After all, anything new involving technology is expensive, until scale production reduces prices.

In a short time, with international competition, photovoltaic solar panels will revolutionise Mato Grosso do Sul, Brazil and the world, because if this is already happening at a rapid pace in Europe, where the sun shines less, what can we say about tropical latitudes? In Brazil, states such as São Paulo, Ceará, Paraíba, Bahia, Tocantins, Rio Grande do Norte and Rio Grande do Sul have taken the lead. And in this race, sometimes they choose to build solar plants, sometimes they choose to set up photovoltaic panel factories, because in addition to taking advantage of incentives from the federal government (BNDS/Climate Fund credit line; Proinfa/ Programme to Encourage Alternative Sources of Electric Energy), as well as from state governments, among them the exemption from ICMS, as is already the case in Mato Grosso do Sul and more than twenty other Brazilian states.

And in the wake of these events, Mato Grosso do Sul is also expected to move in this direction - futuristic and promising, given its potential, since it has the privilege of being geographically situated between the 14th and 27th parallels, which gives it good levels of solar radiation.

According to preliminary studies by the Ministry of Mines and Energy, Mato Grosso do Sul has the greatest potential for solar energy in the Centre West region. That's why, as a new and modern state, it needs to establish a position on this new energy frontier, including the issue in a strategic government policy agenda. An agenda that includes an information centre, tax incentives for businesses, investment in research and technological training.

Rio Grande do Norte, for example, has already signed partnerships with Petrobras and with the National Industrial Apprenticeship System (SENAI) to train technicians at middle and high school level, with a keen interest in increasing its energy potential and generating new jobs, like New Germany.

Mato Grosso do Sul already has a technical study group to deal with the issue, linked to the State Energy Management.

29 HANKLEMAN, J. & MARTIN, C. Solar energy could be cheaper than coal by 2025. Rev. Exame, Ed. Abril, São Paulo, 03.jan.2017.

CHAPTER 5

FISHERIES ENGINEERING AND SUSTAINABLE DEVELOPMENT

Fish meat, which has already become the most widely consumed meat in the world, continues to grow in demand, given the current global demand for high-protein foods. As a result, aquaculture (the commercial rearing of aquatic animals, which includes fish farming), according to Crepaldi et alii (2006), continues to grow faster than all sectors of animal production.[30] He agrees with Lester Brown, for whom aquaculture is currently the fastest-growing sector in the world food economy.[31]

And how auspicious it is to know that, in addition to having an ever-growing domestic market, according to the FAO (2010), Brazil is one of the few countries able to meet the world's growing demand for fish.

The FAO is right, because our country has the greatest water potential on the planet. And with the excellence of the market (an increase, according to the FAO, of 100 million tonnes/year), fish farming can no longer be seen as a secondary, complementary rural activity. As a source of income and social development, it has become a predominant activity.

In this context, Dourados and the region, which is one of the most favourable areas for fish production, has fortunately become aware of this vocation. Here, the abundance of water resources is visible. And these resources, combined with one of the country's gentlest reliefs, have ended up creating wide-open valleys, with an incredible wealth of varjões - privileged spaces for fish farming, given the ease with which they can be used to set up so-called dugout ponds. But the richness of the rivers is also inviting for the installation of net ponds. And the Paraná River itself is available for this purpose. With its current configuration, giving way to huge artificial lakes, it couldn't offer better conditions for this activity.

Since the beginning of the century, Dourados has had plans to expand local and regional fish farming. In partnership with the federal government, the municipality already has a Fish

30 CREPALDI, DV. et alii. Production System in Fish Farming. Revista Brasileira de Reprodução Animal, v.30, n.3-4, p.86-99, 2006.

31 BROWN, L. F. Fish Farms. Wolrdwatch.org website and www.wwilma.org.br. Available at http:/woldwastc.org.br/articles/012.html.

Farming Research Laboratory at the Embrapa unit (which, by the way, in early 2013 received 1.2 million for improvements).

It also has a Fish Farmers' Cooperative, support for the Fish Festival and will soon have a Fish Refrigerator (in the final stages). To crown such an important production chain, Dourados also has the long-dreamed-of Fisheries and Aquaculture Engineering course at the Federal University of Grande Dourados - UFGD.

With all this background, and with Dourados being the Brazilian city chosen to host the 1st Brazilian Congress on Native Freshwater Fish Production in 2012, there is no doubt about the success of this activity in this municipality and region.

In addition to boosting local and regional development, especially because of its capacity to generate new rural and urban jobs, fish production is also an environmentally sustainable activity. Where there are controlled ponds, for example, the impounded water is released slowly into the rivers, which helps to control floods and ebb tides, and where there are ponds, you can't even think about pesticides.

As every successful activity is the result of planning and research, imagine the potential of Dourados with the Fisheries Engineering Course, integrated with Embrapa's Fish Research Centre. There will certainly be improvements in water quality, species genetics and reproduction, feed quality, among many others, which will effectively result in lower costs, better fish quality and greater sustainability.

And since fish is one of the most nutritious and healthy products, and the region enjoys other advantages, such as transport logistics (paved roads), abundant raw materials (grain for feed), promising markets (internal and external), modern technology (in progress), added to the organisation of fish farmers, it can be taken for granted that Dourados and the region will become a benchmark in this important field of activity.

ENVIRONMENTAL LIABILITIES AND THE RACE AGAINST TIME

Environmental liabilities are considered to be all the actions related to the most diverse activities that result in environmental damage and losses and generate obligations for reparation. Due to the importance that has been given to ecological science, especially since the 1970s, this concept has been given generous space in the press.

Issues such as environmental damage and the demands for its reparation are coming to the fore today as never before. This is a concern that has become universal. It has also become clear that, in recent decades, technological advances and the expansion of economic activities have resulted in unprecedented pressure on the environment.

The result is environmental liabilities that are present in almost every geographical area and mark significant ecological losses. Losses that jeopardise the future of life itself.

In Mato Grosso do Sul, for example, environmental liabilities can be found in water pollution, silting up of rivers, obstruction of springs, open-air urban rubbish dumps, the cumulative effect of pesticides and illegal deforestation, among many others. And if each liability is already a problem, imagine the sum of them, knowing that the effects generated by one end up adding to the effects of others, however far apart they may be. The result is the emergence of gigantic liabilities that end up undermining nature to its most precious foundations.

From this perspective, burning and deforestation at a local level end up fuelling planetary liabilities such as the greenhouse effect (global warming), ozone depletion, melting glaciers, rising sea levels and the impoverishment of the biosphere. Pollution from rivers reaches the sea, adding to its acidity and fuelling the extinction of marine species. Despite this, the false premise that the ecological issue is an obstacle to economic development still echoes.

Fortunately, there is growing awareness that enterprises, whether agricultural, commercial or industrial, are directly responsible for their environmental liabilities because they are integrated into the environment. For this reason - and because of the delicate situation in which the environment finds itself - there has been growing concern about ecological efficiency. Along these lines, some companies have adopted environmental management not just as a legal obligation, but because they want to incorporate sustainable development as an integral part of their business, with a view to a better future for the human family.

In fact, modern managers, in tune with our times, know how important prevention is. They know that the accumulation of environmental liabilities over months and years, in addition to damaging society and nature, can generate costs in terms of repairing damage, fines, compensation and, in some cases, huge marketing expenses to restore the company's image in the eyes of an increasingly demanding society concerned about sustainability.

The environmental race is already against time. Developments really need to be connected with nature. After all, it is by preventing and/or eliminating environmental liabilities that the future desired by all is being projected.

TRANSITION FROM AGROCHEMISTRY TO AGROECOLOGY

The environmental debate, which is already present in all social activities, is particularly important when it comes to farming. And it is through agroecology that many have come to see agriculture as an ecological activity.

Although ancient, agroecology was only systematised as a science in the 1980s. Recently, however, it has taken on an air of novelty by presenting itself as an alternative to the dominant agrochemical model. The proposal has been for biodiverse and clean agriculture. This science, imbued with the principles of environmental and social sustainability, has been gaining public recognition to the point where we can already talk about the beginning of a transition from agrochemistry to agroecology.

Agroecology aims to work in partnership with nature, not against it. It treats the land as a living being, based on the responsible use of natural resources. Furthermore, this science has not come to overthrow local knowledge, accumulated by peasants for centuries, but rather to value and qualify it, given that peasant knowledge and scientific knowledge go hand in hand and complement each other. Nor to conflict with conventional producers, but to present them with the good news, so that they too can become part of this transition.

FAO studies have already shown the potential of clean food production. By prioritising quality, Agroecology, with its natural technological base, has shown how much it is possible to get clean, healthy food from the land that prevents and even treats diseases, thus guaranteeing food security.

Free from pesticides and chemical fertilisers, agroecological production of fruit, vegetables, cereals and fibres has more concentrated nutrients and is certainly much more pleasant in aroma and taste, and much richer in vitamins, proteins and minerals. This means that people eat less and eat more, without having to pay more for doctors, medicines and environmental restoration.

Through agrochemistry, people end up eating more to satisfy their hunger, because agrochemicals, by eliminating soil microorganisms, worsen soil fertility. Although beautiful in appearance, the products end up empty, poor in nutrients, because the mineral and vitamin content drops sharply.

On the other hand, society needs to know that it is swallowing poison. And the amount is not small. More than 5.2 litres of pesticides per year are being ingested per person in Brazil. It sounds incredible, but it's true. More than a million tonnes of poisons a year are being applied to Brazilian crops - a fact that unfortunately puts our country in first place in the world consumption of pesticides.

According to the World Health Organisation (WHO/UN), pesticide poisoning already affects three million people a year worldwide. And in Brazil, according to the National Health Surveillance Agency (ANVISA), poisoning and deaths from pesticides have already become a serious public health problem. Also, according to the literature on the subject, it can be seen that the formula of these poisons can cause a multitude of pathologies, such as sterility, cataracts, allergies, neurological, respiratory, cardiac, immune and endocrine disorders; reduced attention, memory and motor coordination; blindness, dizziness, nausea, depression; nerve, liver and kidney damage; foetal malformations, among other problems, such as cancer.

The search for clean food is gaining in importance to the point that it is being repeatedly recommended by the WHO, as it reduces the rate of illnesses resulting from 'modern' eating patterns. Indeed, it cannot be accepted that widespread poisoning is the necessary evil to feed the population, since the production of food without poisons and chemical fertilisers is no longer taboo. We must recognise the importance of this transition, which is being proposed worldwide, given the growing demand for clean, organic products, motivated by the desire to have a source of health at the table.

In Brazil there has been a gradual increase in sales of agroecological production in recent years, a fact that signals greater attention to the laws of life, because agrochemicals will always contaminate rivers, springs, groundwater, fauna, flora and everything else. And, worse, contaminate food and jeopardise people's health.

The application of pesticides needs to be rethought, for the sake of public health and the entire biological base. Furthermore, prestigious world organisations are already calling for their complete extinction. From this perspective, agroecology opens up the way to producing rich, healthy food, generating more jobs and income in the countryside, expanding genetic heritage, biodiversity and life itself.

Fortunately, our country has already woken up to this growing cycle of renewal, through which it will be able to conquer a market that is even being opened up worldwide, with ample possibilities for Brazil to become a global benchmark in agroecological production, given the continental dimensions of its territory and the diversity of its soil and climate.

CHAPTER 6

CREOLE SEEDS: BIODIVERSITY AND MONOPOLY

"Many small people, in small places, doing small things, will change the face of the earth" (African proverb)

Creole seeds are natural seeds that are resistant to pests and diseases and have not been genetically modified, as is the case with the large volume of seeds produced today. They are seeds that have been widely adapted to the production sites and improved by peasant families since their distant ancestors. This marvellous genetic material, which is healthy and rich in nutrients and which forms the basis of agroecology, is a historical legacy dating back more than 10,000 years and is now the heritage of generations, in other words, of humanity.

Today, despite the advances in so-called biotechnological and modern agriculture, it is clear that monocultures are taking over the countryside and, with the current model imposed by the monopoly of the large transnational companies, which are fighting for control of food production worldwide, humanity is losing a lot. These companies already control much of what we eat and how much we pay. Researchers have revealed that there are more than 10,000 species of edible plants and that ancient agriculture used more than 500 species. But with the rise of the multinationals, the consequences are there - in just over 20 years, humanity has seen a drastic reduction in seed varieties, and many have already been lost.

In fact, conventional (industrial) agriculture has reduced the basis of our diet to less than ten species, which are the ones that make the most profit on the market. Wheat, rice, corn and soya, for example, account for almost all of the world's grain consumption.

In fact, the agrochemical and monopolising model applied to the countryside has disregarded the ecological practices of peasant agriculture and, since the so-called Green *Revolution*, creole varieties have been criticised and even seen as backward. They disqualify a wealth of biodiversity in order to impose their own seeds, such as transgenic seeds, which are highly dependent on chemical inputs and fertilisers and have high production costs. And what's worse: genetically modified.

Traditional or creole seeds are an agroecological alternative, a universal asset that should never fall prey to privatisation. It should be emphasised that nowadays, when farmers buy a modified seed, they are obliged to sign a contract prohibiting its replanting, commercialisation and exchange. And if this is already done with commodities, imagine the risk of them advancing on creole seeds and wiping peasants off the map. In Brazil, however, legislation was passed in 2003 to authorise the use of creole seeds, fortunately.

In recent years, family farmers have shown their indignation. All over the world they are reacting. Via Campesina International, for example, has launched the campaign: Heritage of the Peoples at the Service of Humanity, with the objectives of preserving healthy seeds, combating the manipulation, monopolisation and imposition of transgenic seeds. As you can see, peasants are no longer willing to lose control over their main input - the seed! And despite the rapid advance of modern technologies such as biotechnology, the demand for creole cultivars is increasing like never before. As a result, the rescue and multiplication of these seeds, also known as seeds of hope, solidarity and passion, has been growing in Brazil and around the world. Thousands of family farmers have already opened their eyes and many of these seeds have been rescued, multiplied and exchanged in opposition to the monopolies. With this, they are once again seeking their roots and their independence, as an appeal for the right to sow, to pass on to their children and grandchildren the secrets of how to save seeds, how to value ancestral wisdom, to produce food with ethics and quality and free from multinationals.

They do make partnerships, but with technicians and institutions committed to the cause. As a result, thousands of properties in Brazil already have organic certification, and organic consumption in the country is growing by 40 per cent a year. What's more, thousands of tonnes of creole seeds have been produced in recent years.

In Santa Catarina, just one Creole Seed Processing Unit (UBS) has already delivered seeds to more than 100,000 families in the three southern states, which means not only the rescue of hundreds of varieties, but also peasant virtue in the face of their cry for independence for a future of socio-environmental balance.

Modern agriculture, with all the technology at its disposal, is in fact centred on a small number of crops. In Brazil, the processed products in supermarkets are almost entirely based on five crops: rice, corn, wheat and soya, which is worsening agricultural biodiversity (agrodiversity), with serious repercussions on environmental instability and, worse, nutritional deficiency, as it is known that hybrid seeds are increasingly less adapted and more impoverished.

The fact that small family farmers are ardently seeking to revalorise creole seeds is not a step backwards. On the contrary, in addition to local biodiversity, they want to preserve genetic wealth with seeds that are more resistant and less dependent on inputs. They want to guarantee sustainability based on agroecology. They want to produce healthy food, and above all, they want to help stop something serious - the extinction of Creole seeds. According to the FAO/UN, it is estimated that in the last century around 75 per cent of all genetic diversity on the planet has disappeared - a process that has intensified since multinationals began to control seeds around the world, jeopardising people's sacred right to freedom of choice.

There is no longer any doubt about the vertiginous decline of creole seeds, especially since the Green Revolution and the advance of the production and commercialisation of hybrid and transgenic seeds. That's why creole seeds have become so important that the UN itself has shown how crucial it is to recover and conserve them.

In most Brazilian states, as well as in various parts of the world, there has been an encouraging trend towards fairs for creole seeds and agroecological products, which bring together thousands of interested people. In Mato Grosso do Sul, one of the country's main agro-export frontiers, in the town of Juti, the Pastoral Land Commission and a group of farmers accepted the challenge and the Creole Seed Fairs are now in their 13th year, welcoming peasants and other participants from various regions of the state and country. In Rio Grande do Sul, the Canguçu Fair brings together more than 20,000 people.

They are fairs that provide full conditions for the exchange and propagation of these seeds, as well as for sharing information and articulating resistance. The programme includes talks and seminars on a variety of topics, such as biodiversity, agroecology, environmental recovery, the harms of pesticides, the history of seeds, rescue, conservation, organic products, certification, the use and preservation of biomes, traditional techniques, ethical markets, the fight against seed monopolies and peasant culture. From start to finish, they emphasise the defence of this essential heritage for humanity, as well as an agroecological and sustainable agricultural model that rebuilds the genetic base, restores food sovereignty to the people, promotes regional culture and opposes the imposing and dominant agricultural model.

MATO GROSSO DO SUL AND SUSTAINABILITY - IN CONCLUSION

Intelligent life on the planet hasn't been all that intelligent. Over time, human beings continue to have a negative impact on ecosystems, with visibly disastrous results almost everywhere. The major world conferences on the subject have demonstrated how much the human species depends directly on healthy ecosystems. Today, environmental issues are global and conservation has become essential.

Mato Grosso do Sul, with its gentle topography, exuberant vegetation, large tracts of fertile soil, prodigious hydrography and well located in the interior of Brazil and South America, just had to be hit by the accelerated expansion of agricultural fronts, to the point of rapidly integrating itself into the globalisation route. However, it is clear that it has fallen far short of making sustainable use of its natural resources. In this state, the incredible loss of biodiversity is already an eye-opener. The absolute majority of biomes are already degraded or under severe pressure. The prevalence of the developmentalist economic model has long been extrapolating ecosystem limits, through unlimited exploitation of spaces, without taking into account the vulnerability and exhaustion of the environment.

Even though measures of proven sustainability are being implemented, there is no denying that others, contrary to the fundamentals of nature and life, are also being implemented, such as the use of pesticides banned in other countries; the illegal and unbridled suppression of centuries-old biomes, including riparian forests; the millions of hectares that have been degraded by erosion, through livestock farming that is still largely absentee; the hypermonocultures, which are increasingly dependent on agrochemicals; the municipality of Corumbá, which continues to lead the national ranking for fires; the fauna and flora that are increasingly confined to tiny forest fragments.

And in the wake of the destruction, there is also the silting up of rivers, streams and lakes, the obstruction and death of springs, water pollution, the depletion of fishing resources, illegal charcoal works, etc. And if the bills are not being collected institutionally, as they should be, nature itself is taking care of it, through the relentless accumulation of environmental liabilities.

After such a significant awakening of environmental awareness, especially after Rio+20, the prevailing developmentalism in Mato Grosso do Sul, Brazil and the world no longer makes sense. The degradation and depletion of natural resources - the great challenges of the current century - call for a transition to an agricultural model that promotes the economy and food production, but which evolves, while there is still time, towards sustainable development.

In the immensity of the territory of Mato Grosso do Sul, if there are still generous spaces for production, there must also be a minimum for preservation. At the very least, environmental laws must be complied with.

In this state, while environmental issues are still not prioritised, the movement in favour of sustainable development remains fragmented and disjointed, despite the range of highly competent environmentalists present here.

The creation of interactive forums for these professionals would certainly enable a new pact with nature to be established in the south of Mato Grosso.

Strictly speaking, they are professionals who are fully qualified to design a pact focused on a holistic, regenerative vision and in tune with restorative measures for pastures, riparian forests, reforestation with biodiversity, the implementation of ecological corridors, the fight against illegal deforestation, the gradual replacement of fossil fuels with clean energies (solar, bioenergy), and with environmental education, agroecology, among others.

Therefore, in these Guaicurus plagues, as well as throughout the national territory, reversing the worsening environmental situation means taking on the noble mission of building the sustainable future of this and new generations.

REFERENCES

1. ALENCASTRO, Catarina. Brazil is first in the ranking of *forest* clearing. O Globo newspaper, 24.09.2014 - updated on 24.09.2016.
2. ALMOAS, Jorge. *IBAMA imposed R$ 51 million in fines in Mato Grosso do Sul in 2010.* Campo Grande News, 21 December 2010.
3. CERQUEIRA, Wagner. *Deforestation in Brazil.* Available at mundoeducacão.uol.com.br/geografia/desmatamento-no-brasil.html.
4. MORELLO, Thiago Fonseca. *Deforestation is not development.* Planeta Magazine, São Paulo, 14 December 2012.

5. *Production should grow using degraded areas*. Interview with Eduardo Riedel for Correio do Estado. Campo Grande, MS, 30.Sep.2011.

6. BARRETO, Paulo. *Deforestation is unnecessary.* Folha de São Paulo - Opinion, 27 June 2009.

7. *Riparian Forests Programme for the State of Paraná.* Available at htp/www.meioambientre.pr.gov.br/modules/conteúdo/conteúdo phd?con

8. INPE News, 05.Jun.2012, Available at: HTTP:/www.cpetec.inpe.br/noticias/noticias/128679

9. *Atlas of Atlantic Forest Remnants, results by municipality.* SOS Mata Atlântica Foundation, May 2009.

10. Cavalcante, Bárbara. *Mato Grosso do Sul has the highest number of fires in the country.* Correio do Estado - Weather. Campo Grande - MS, 12 January 2017.

11. TEIXEIRA, M.A. *Agricultural changes in Mato Grosso do Sul.* Doctoral thesis. São Paulo, 1989.

12. PIRES, W. *Degraded pastures? What should I do: reform or recover?* Circuito da Pecuária, 12.ag.2016. Available at http:/www.circuitodapecuaria.com.br/

13. KICHEL, A.N. Half of Mato Grosso do Sul's pastures are degraded, says EMBRAPA. Interview with Paula Vitorino, Campo Grande News- Rural. Campo Grande-MS, 05.ag.2011.

14. MACEDO, M.C.M, et alii. Pasture Degradation, Alternatives, Recovery, Renovation and Forms of Mitigation. EMBRAPA, Campo Grande-MS, 1993.

15. SILVEIRA, E. The *advance of deserts.* Rev. Planeta - Meio Ambiente. Ed. 519. Cajamar-SP, 21 January 2016.

16. CALIXTO, Bruno. *Scientists estimate how many people live with a lack of water in the world: 4 billion.* Época Magazine, Rio de Janeiro, 01.nov.2016.

17. National Council for Food and Nutrition Security. *Studies show a link between pesticide consumption and cancer.* Source: Ascom/Cosea, Brasília-DF, 08.Apr.2016.

18. MANSUR, Alexandre. *The Cerrado has lost a Rio de Janeiro in ten years. Find out the consequences.* Blog Planeta Época. Rio de Janeiro, 01.nov.2016. Available at

http://epoca.globo./ciencia-e-meio-ambiente/blog- do-planeta/

19. *With almost 20,000 springs, the devastation of Cerrado basins impacts the lives of 88.6 million people.* Ecodebate - News, by Redação. 25.Apr.2011

20. VARGAS, R. In Mato Grosso do Sul, 270,000 hectares turned into coal in three years. Folha de São Paulo - Science. São Paulo, 05.Apr.2010.

21. STURZA, C. MS will become the country's largest pulp producer from 2017. Midiamax, Campo Grande- MS, 18.ag.2016.

22. Police close charcoal factory in MS and seize 181 cubic metres of firewood and charcoal. Globo.com - G1 MS, Campo Grande-MS, 08.jan.2017.

23. Rescuing the Cerrado from Extinction. Ecoa, Campo Grande MS, 07.Dec.2016 - Available at riosvivos.org.br/notícias- ecoa/resgatar-o-cerrado-da-extincao.

24. LIMA, J. Plantas e Dor. Contribution to the Anthropological Study of Pain Treatment. Almada, University of Coimbra, 2010.

25. Costa, Luciano Martins. Observatório da Imprensa. Ed. 410, 05/Dec/2016. Available at URL of article: www.observatoriodaprensa.com.br

26. Florios, Daia. I will do good to the Earth. Greenme.com.br. 09/Nov./2015. Available at https://www.facebook.com.

27. MATTAR, H. Vitae Civilis discusses transition to a green economy. Conference Ethos,11.06.2012. Available at at siteuniethos.org.br.ci2012/?=888

28. HANKLEMAN, J. & MARTIN, C. Solar energy could be cheaper than coal by 2025. Rev. Exame, Ed. Abril, São Paulo, 03jan2017.

29. CREPALDI, DV. et alii. Production System in Fish Farming. Revista Brasileira de Reprodução Animal, v.30, n.3-4, p.86-99, 2006.

30. BROWN, L. F. Fish Farms. Wolrdwatch.org website and www.wwilma.org.br. Available at http:/woldwastc.org.br/articles/012.html.

MIX
Papier aus verantwortungsvollen Quellen
Paper from responsible sources
FSC® C105338

Printed by Books on Demand GmbH, Norderstedt / Germany